A Laboratory Manual for
INTRODUCTORY OCEANOGRAPHY

Interpretations and Applications

Lawrence A. Wiedman
Monmouth College

West Publishing Company
St. Paul New York Los Angeles San Francisco

COPYRIGHT © 1992 by   WEST PUBLISHING CO.
                     50 W. Kellogg Boulevard
                     P.O. Box 64526
                     St. Paul, MN 55164-1003

All rights reserved
Printed in the United States of America
99 98 97 96 95 94 93 92 8 7 6 5 4 3 2 1 0

ISBN 0-314-00017-8

CONTENTS

PREFACE.. i

LABORATORY ONE - WORLD MAP PROJECTIONS
    WORLD MAPS AND MAP PROJECTIONS........................ 1
    EXERCISES.. 13
    GLOSSARY OF NEW TERMS................................. 21
    FURTHER READINGS...................................... 21

LABORATORY TWO - ORIGIN OF OCEAN BASINS
    MARINE AND CONTINENTAL ROCK DIFFERENTIATION........... 23
    EXERCISES.. 30
    GLOSSARY OF NEW TERMS................................. 33
    FURTHER READINGS...................................... 34

LABORATORY THREE - MARINE SEDIMENTS
    AND ROCKS FORMED FROM THEM............................ 35
    EXERCISES.. 38
    GLOSSARY OF NEW TERMS................................. 46
    FURTHER READINGS...................................... 46

LABORATORY FOUR - PLATE TECTONICS
    PLATE TECTONICS AND COASTAL DEVELOPMENT............... 47
    EXERCISES.. 52
    GLOSSARY OF NEW TERMS................................. 60
    FURTHER READINGS...................................... 60

LABORATORY FIVE - BATHYMETRY
    TOPOGRAPHIC MAPS AND BATHYMETRIC CHARTS............... 61
    MAKING A BATHYMETRIC CHART............................ 65
    TOPOGRAPHIC AND BATHYMETRIC PROFILES.................. 69
    EXERCISES.. 71
    GLOSSARY OF NEW TERMS................................. 79

FURTHER READINGS.. 79

LABORATORY SIX - COASTAL FORMATION
COASTLINES AND COASTAL FORMATION.......................... 81
EXERCISES.. 83
GLOSSARY OF NEW TERMS...................................... 87
FURTHER READINGS... 87

LABORATORY SEVEN - MARINE CHEMISTRY
DISSOLVED SOLIDS AND SALINITY DETERMINATION............... 89
EXERCISES.. 92
GLOSSARY OF NEW TERMS...................................... 96
FURTHER READINGS... 96

LABORATORY EIGHT - WAVES
MEASURING WAVE PARAMETERS.................................. 97
EXERCISES.. 104
GLOSSARY OF NEW TERMS...................................... 106
FURTHER READINGS... 107

LABORATORY NINE - BEACH FORMATION
BEACH FORMATION AND LONGSHORE DRIFT....................... 109
DEMONSTRATIONS/EXERCISES................................... 109
QUESTIONS FOR DISCUSSION................................... 117
GLOSSARY OF NEW TERMS...................................... 118
FURTHER READINGS... 118

LABORATORY TEN - SEA LEVEL AS A DATUM
SEA LEVEL AS A REFERENCE POINT............................. 119
EXERCISES.. 123
GLOSSARY OF NEW TERMS...................................... 129
FURTHER READINGS... 130

LABORATORY ELEVEN - DELTAS
DELTAS AND DELTA FORMATION................................. 131
LABORATORY DISCUSSION...................................... 134
DEMONSTRATIONS/EXERCISES................................... 135
GLOSSARY OF NEW TERMS...................................... 138
FURTHER READINGS... 138

LABORATORY TWELVE - BEACH SAND
BEACH SAND DIVERSITY....................................... 139
EXERCISES.. 140
GLOSSARY OF NEW TERMS...................................... 144
FURTHER READINGS... 145

LABORATORY THIRTEEN - MARINE MICROFOSSILS
FORAMINIFERA, SPONGES, BRYOZOANS, ETC...................... 147
EXERCISES.. 150
GLOSSARY OF NEW TERMS...................................... 156
FURTHER READINGS... 156

LABORATORY FOURTEEN - POINT COUNTING
STATISTICAL EVALUATION OF BEACH MATERIAL................... 157
DICHOTOMOUS KEYS... 160

```
    EXERCISES..................................................  160
    GLOSSARY OF NEW TERMS......................................  167
    FURTHER READINGS...........................................  167
```

LABORATORY FIFTEEN - CAMINALCULES
```
    DETERMINING BIOLOGICAL DIVERSITY AND PHYLOGENY.............  169
    EXERCISES..................................................  169
    GLOSSARY OF NEW TERMS......................................  171
    FURTHER READINGS...........................................  179
```

LABORATORY SIXTEEN - INVERTEBRATE ECOLOGY
```
    BIVALVE ECOLOGY AND PALEOECOLOGY...........................  181
    EXERCISES..................................................  186
    GLOSSARY OF NEW TERMS......................................  188
    FURTHER READINGS...........................................  189
```

LABORATORY SEVENTEEN - MARINE ECOLOGY
```
    FISH AQUARIUM STUDY........................................  191
    EXERCISES..................................................  193
    FURTHER READINGS...........................................  195
```

TRITON'S REVENGE - WHO RULES THE SEA?
```
    A SIMULATION ON THE LAW OF THE SEA CONFERENCE..............  197
    SCENARIO...................................................  199
    BRIEFS ON COUNTRIES AT THE LAW OF THE SEA CONFERENCE.......  200
    TALLY SHEET TO SUMMARIZE DATA..............................  201
```

PREFACE

Before you begin the laboratory portion of your course, I would like to have a few moments of your time. I would first like to welcome you to what I hope you will find an exciting and rewarding class and laboratory experience. The labs were designed with a few premises you may appreciate hearing about.

One of the main frustrations voiced by many oceanography instructors over the past few years has been that it is very difficult to teach a meaningful laboratory with the class unless, of course, their institution is near an ocean. As a response to those frustrations, I have written the lab text you are about to use. It is intended to be used principally by undergraduate, non-majors, most of whom are using the course to satisfy the general education components of their degree at institutions not near oceans. It is further designed that, whenever possible, the skills being taught will be practiced in a "hands-on" manner. Contrary to some other courses where you are expected to utilize external printed references from the library, these labs have been written so that the questions are thought through, not looked up! You are encouraged to utilize the experiences of the whole class; converse with one another. At this level, you each bring into the course a rich diversity of experiences; use them.

The labs try to incorporate life skills; I have tried to concentrate on topics that fit the premise of being "news you can really use". I hope you will derive from the labs an appreciation of how working scientists approach problems. Prior to class time, you are expected to read the material. If you do not, you may count on being somewhat lost. During the first 15-25 minutes, your instructor will demonstrate any specific procedures needed to finish the lab. The bulk of the lab time will be devoted to your completing the lab exercises. The last 20-30 minutes are set aside for discussion, hopefully in a seminar fashion, where each of you are prepared to input your observations from the lab. During the lab, write down your observations as directed. These may be collected and graded, or they may provide the basis for your input to the discussion, depending on the method your instructor chooses.

As a closing note to this preface, let me invite you to **have fun!!!** Science has always been fun for me, and I cringe when I hear people talk about the horrible experiences they have had in required science courses. Nearly all the scientists I know became scientists because they truly enjoyed the thrill of discovery and the excitement of experimentation. Enjoy your class, participate enthusiastically, and, if you don't watch out, you'll learn something.

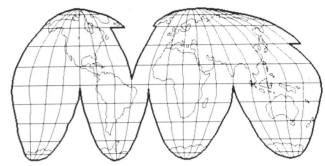

There is something in being near the sea, like the confines of eternity. It is a new element, a pure abstraction. The mind loves to hover on that which is endless, and forever the same... I wonder at the sea... that vast Leviathan, rolled round the earth smiling in its sleep, waked into fury, fathomless, boundless, a huge world of water-drops - Whence is it, whither goes it, is it of eternity or of nothing?

Willian Hazlitt, Notes of a Journey through France and Italy, 1826

Laboratory ONE

World Map Projections

World Maps and Map Projections

"Where in the world are we?" is a frequent concern of many of us. Perhaps we only utter it rhetorically, but it is a question worth consideration. Before you go further in your readings, turn to the beginning of the exercises for this chapter and pull out the page marked for exercise 1, and place it before you. Now draw a map of the world without consulting a reference. This assignment is essential to the lab; so don't bypass it. Do the best you can at filling in the landmasses with political names. Include a line to represent the equator. After you have finished, set the drawing aside and continue reading. Past classes have found it helpful to tack up these map versions on the walls allowing the map makers to remain nameless.

Maps and charts are critical and essential tools for any study of the earth, and certainly oceanography fits this description. Few of us realize all the choices we have available to us when we wish to find our position on the earth. Without a doubt, a globe, especially one with raised relief areas representing the mountains and depressed areas representing basins of the terrestrial and submarine areas of the world, is our best choice. However, globes are frequently not practical to carry with us or tack on the wall. We are then left with the next best choice, a flat world map.

When choosing a map several factors need to be considered: scale, color, size, and price. Few of us include among this list perhaps the most important factor: the **map projection**. One problem inherent in representing a three dimensional object, especially a sphere, on a flat surface is distortion of the view. To see how much distortion, let us take a grape and smash it on a table with a flat board. The resulting mess certainly has little resemblance to a grape. This is akin to the fate of the world when flattened onto a planar surface.

Although several very clever cartographers (map makers) have addressed this problem in the past, they have, to date, discovered no method that is distortion free. Each method has a built in set of compromises which must be addressed. Among the compromises are:

1. **Equality of Area** such that a user may quickly and accurately compare the true relative size of one area to another; for example, the true area of Greenland to that of Australia;

2. **Fidelity of Axis** meaning that most people want north to be at the top of the map (although there is really no valid scientific reason why north ought to get such a "high" priority);

3. **Fidelity of Position**, the ability to have North-South lines intersect East-West lines at 90 degree angles and East-West lines (latitudes) run parallel;

4. **Equity of Linear Scale** such that any two points A and B on the map would appear to be the same distance apart no matter where they were positioned on the map;

5. **Accuracy of Shapes**, Africa ought to really look like Africa;

6. **Fidelity of Angles**, compass points ought to read correctly anywhere on the map.

Other factors could be included as well on such a list, for instance, the **location of the equator**. Most of you probably thought that on all world maps the equator was placed in the center where it ought to be. Look closely. On most world maps this is not the case. Why not?

We repeat that no ideal projection of the earth has ever been devised. **Cartographers** are limited in choices. Some of the ideal qualities of a flat projection mentioned earlier are even considered mutually exclusive, i.e., you cannot have one if you have another.

The most frequently used map of the world, and the one most of you probably conjured in your minds when asked to draw the world, was designed in 1569 as an aid to European navigators. It is named for its creator, Gerhard Kremer, and is known as the Mercator Map (merchant is the English translation of Kremer, and Mercator is the latinized version of the English word merchant). (Figure 1-1). When this map hit the market it was an instant success. It was great for navigators but has several limitations for other users.

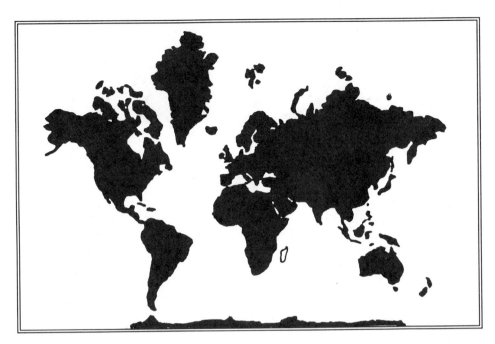

Figure 1-1. The familiar Mercator map, a scientific breakthrough in 1569, but outdated today. (From *A New View of the World*, by Ward Kaiser, copyright 1983 by Friendship Press, 475 Riverside Drive, New York, NY 10115.)

The Mercator map was designed by a process that has been described as similar to placing a light in the center of a transparent globe and projecting the earth's features onto the inside of a paper cylinder wrapped around it. When the cylinder is unrolled, the map features are on the flat paper. See figure 1-2A.

Other projections view the world as if the viewer was at some point above the earth. Obviously, only one hemisphere can be seen at a time using these methods. Two such examples are shown in the Lambert and Polyconic Projections (Figure 1-2B).

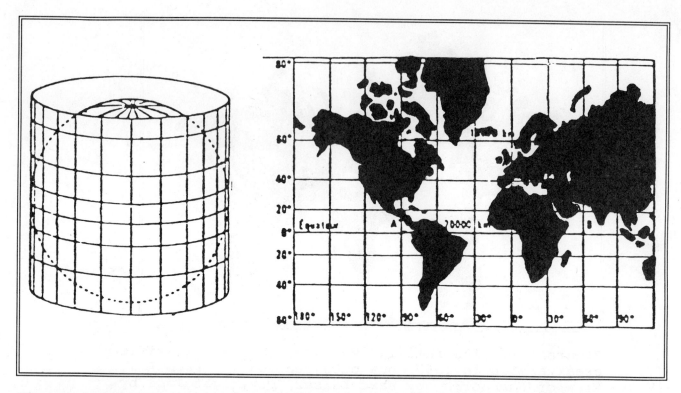

Figure 1-2A. To get a rough picture of the projection method used in the Mercator map, imagine a light bulb inside a transparent earth, projecting the earth's features into the inside of cylinder wrapped around it. When the cylinder is unrolled, the sphere's features are displayed on a flat surface. (From *A New View of the World*, by Ward Kaiser, copyright 1983 by Friendship Press, 475 Riverside Drive, New York, NY 10115.)

The problems with this map are many, but remember that there will be problems with all flat projections of the world. The Mercator map progressively distorts shapes away from the equator, i.e. things at the poles look much larger than they really are. This helps explain why Greenland, an island, looks so large compared to Australia, a continent. Other distortions appear as well, and over prolonged viewing, have given early elementary school children all the way to advanced degree students an incorrect view of the world. Most Mercator maps place the equator about two thirds of the way down from the top. Africa and other southern hemisphere areas are dramatically slighted. Yet this is still, by far, the greatest selling map of the world. Perhaps that shows you who buys maps. Or, this pattern may simply be indicative of the fact that most continental land area is in the Northern Hemisphere.

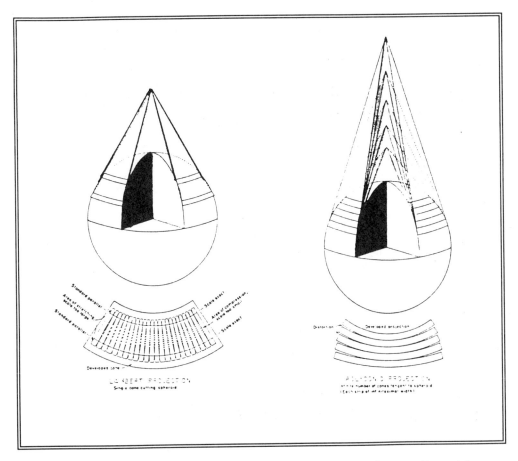

Figure 1-2B. Other methods of projection, for the Lambert and the Polyconic Projections are shown here. From Moris M. Thompson. *Maps for America.* United States Geological Survey, Third Ed., 1988.

Other projections are available. In 1904, American engineer Alphous Van der Grinten developed a map that many of us have seen thanks to the publishers of National Geographic Magazine who inserted this map into their December 1981 issue. This map also gives unequal prominence to the Northern Hemisphere and lacks both fidelity of axis and position. Recognizing these problems, the National Geographic Society, in 1988, announced that they were changing from the Van der Grinten Map to the Robinson Projection (Figures 1-3 and 1-4) for all its map inserts and figures.

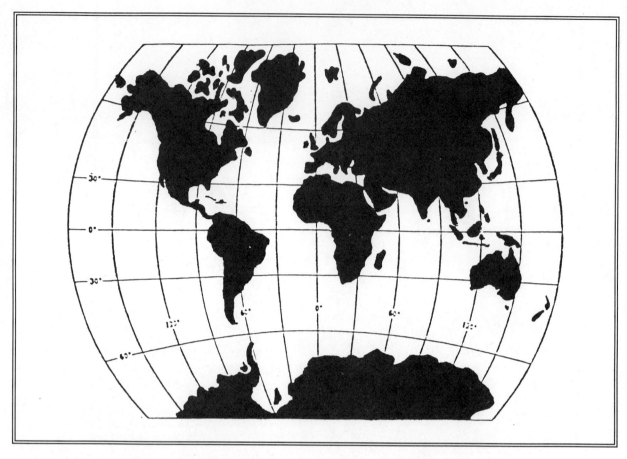

Figure 1-3. The Van der Grinten Projection. (Courtesy of the Geography Department, University of Wisconsin).

This new map is named in honor of Arthur H. Robinson of the University of Wisconsin-Madison, the recognized leader of American cartography and was developed during the 1950's and 60's when Robinson directed the Office of Strategic Services (OSS). This map offers less exaggeration at the high latitudes and more accurate shapes of landmasses, factors that were important to the military complex who funded its production during the cold war.

Others have made attempts to rectify the problems of flat projections. Buckminster Fuller, the creator of the geodesic dome, created his Dymaxion Sky-Ocean World Map in 1927. This ingenious map retains equality of area but lacks a vertical axis and fidelity of position (Figure 1-5).

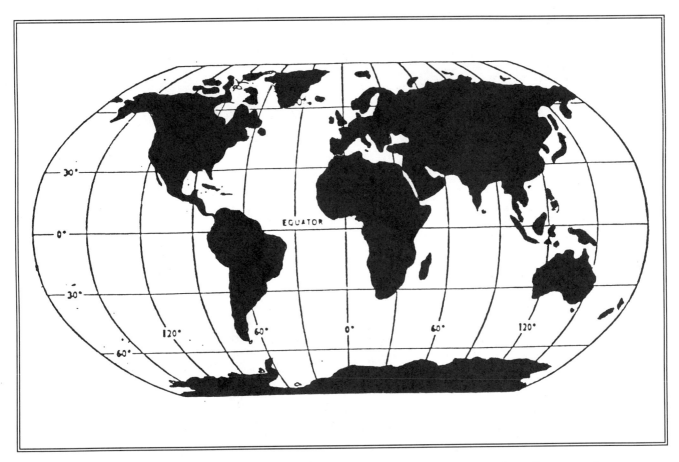

Figure 1-4. The Robinson Projection. (Courtesy of the Geography Department, University of Wisconsin).

This map, although in many ways superior to other attempts to map the world, has not caught on, probably due to its broken surface, sharp angles, and "strange look".

Again, realizing that all flat maps have problems, a new projection was developed in 1974 (published in English in 1983) by Arno Peters, a German historian, and is gaining in popularity (Figure 1-6).

This map would certainly not be the preference of navigators, but actually creates a truer reality of the world in some respects by having fidelity of axis and position and accurately representing all areas according to relative size, i.e. one square inch on the map represents a constant number of square miles. Among the main problems with this projection is that it looks funny and Antarctica is shown as a continuous linear mass. However, some of this is strictly an historical perspective, not one that is inherently incorrect. The South-pole-up map (Figure 1-7) looks funny too, but may actually provide a better view of the earth for some purposes.

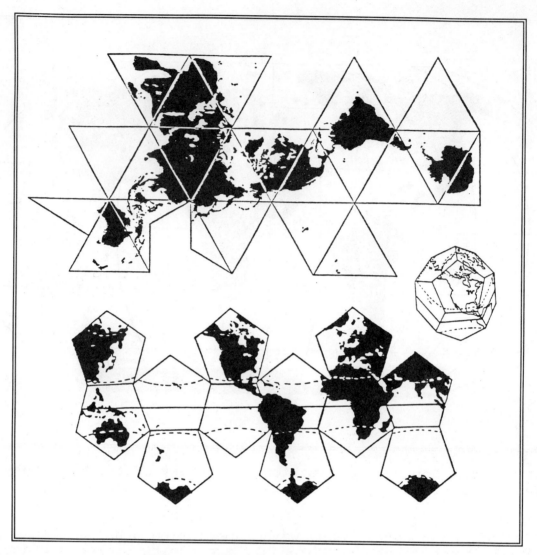

Figure 1-5. Buckminster Fuller's Dymaxion World Map Projection. (Redrawn from B. Fuller, *Ideas and Integrities*, 1969.)

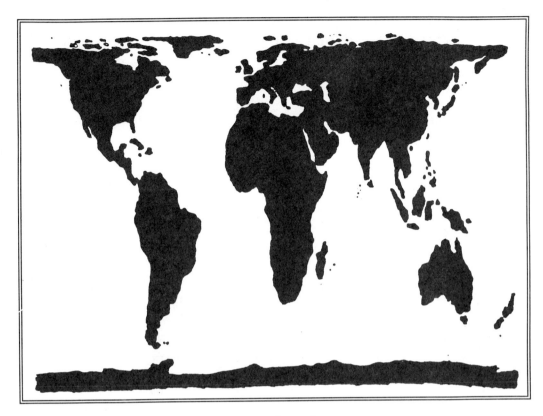

Figure 1-6. The Peters Projection.

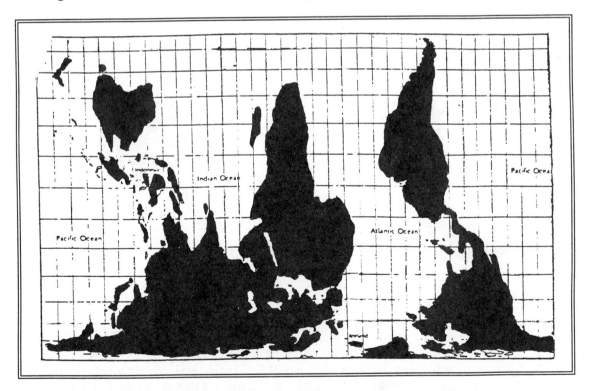

Figure 1-7. Has this map been printed upside down? Is Australia really "down under"? (Both fig. 1-6 and 1-7 from *A New View of the World*, by Ward Kaiser, copyright 1983 by Friendship Press, 475 Riverside Drive, New York, NY 10115.)

Several agencies of the United Nations, as well as the Peace Corps, have adopted the Peters map for use in their publications. We are not attempting to endorse any one map over another; rather we hope to point out the problems and implications of various projections and representations.

Why devote an entire lab exercise to map projections in an oceanography course?

o   First, one must realize that as land masses are distorted by size and shape, so too are the seas.

o   Secondly, the message is clear that by concentrating on the land masses, we frequently forget that the world is really one big ocean with bits and pieces of land strewn about it (Figure 1-8).

o   Additionally, many subtle and not so subtle messages are sent via maps.

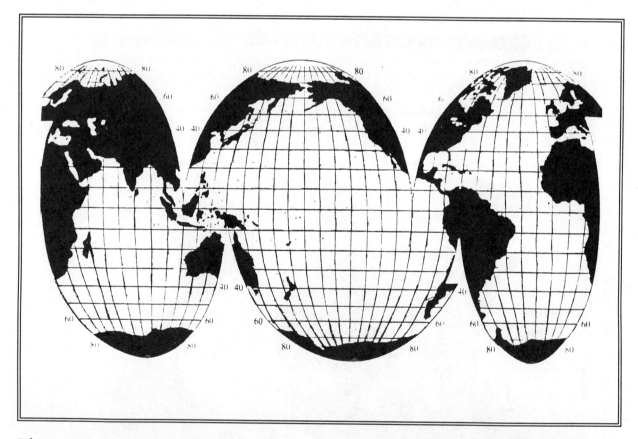

Figure 1-8. A Homolosine World Map Projection. This map view seems to imply that the world is mostly one solid ocean, perhaps a more accurate inference than other maps. (From A New View of the World, **by Ward Kaiser, copyright 1983 by Friendship Press, 475 Riverside Drive, New York, NY 10115.)**

You may have noticed that maps made in the United States tend to place the U.S. in the center of the map; the same is true of Europe for maps made there. One may get a very ethno-centric message that overtly implies greater importance of any one region, especially as geo-political boundaries are being reshaped daily. Ward L. Kaiser in *A New View of the World* believes that the use of the Mercator map, although no fault of Kremer (its originator), has even been a tool to support undue nationalism and perhaps even racism. He writes:

> "I do not assert that this map is the source of the problem but that it lends support to the problem. It does so, not out of any original evil intention, but precisely because many people accept it uncritically. To the extent that people believe it to represent 'truth' or to be 'accurate', to that extent it is inextricably linked to a major, worldwide problem of our time in history.
>
> Ethno-centrism or racism is the belief that one's own people or race are superior to others. The sources of this dangerous disease are many, but surely one of the subtle supports is the pervasive use of a map that, in spite of all known facts:
>
> --enlarges those areas of the world historically inhabited by whites,
> --shifts those same areas to the heart and center of the world's stage, where they do not belong, and
> --minimizes the importance of what we think of as 'the South', including most of what we designate the Third World.
>
> Those who use the Mercator for purposes other than steering a sailboat across the sea may be innocent victims of ethno-centric or racial bias.
>
> How then does it happen that many still use that four-centuries-old map? After all, we constantly confront it in classrooms. And in textbooks. And as background for world news presentations on television. We even see it displayed behind the United States Secretary of State when he interprets foreign policy." (p. 12-13)

This statement by Kaiser certainly is a controversial notion. Do you feel his assertion has merit?

Buckminster Fuller has written in his book *Ideas and Integrities* (1969), "...one of these pictures [a world map] is...the One Ocean World, fringed by the shoreline fragments.... discloses the relative vastness of the Pacific and emphasizes that ocean's longest axis, from Cape Horn to the Aleutians. Oriented about the Antarctic, the waters of the Indian and

Atlantic Oceans open out directly from the Pacific as lesser gulfs of the one ocean."

View each of the maps provided for you thus far. Do any give you a sense of distortion now that you realize how they were drawn? Do all of them? With this in mind, we shall now do a series of exercises designed to re-orient our thinking about the world.

Exercises

1a. **Creating a World Map**
Draw a world map as best you can from memory. Label it however you desire. Place a line on your map to represent the equator.

1b. **After drawing the map...** Estimate from memory or guess, the percent of land and sea on the earth's surface.

Land %	Water %

Now look at your map. What percent does your map show?

Land %	Water %

What is the actual percentage on a globe?

Land %	Water %

Exercise 1.

THE WORLD ACCORDING TO ME

2. **Analyzing "The World According to Me"**
   Return to the maps you and your classmates drew earlier; can you see any of the biases discussed in the text in your map or others' versions of the world? Many people are ashamed of their efforts either because they lack artistic skills or because they lack knowledge about the "realities" of world geography. If you are among these, you are not alone; but ignorance need not be permanent.

Answer to question #2

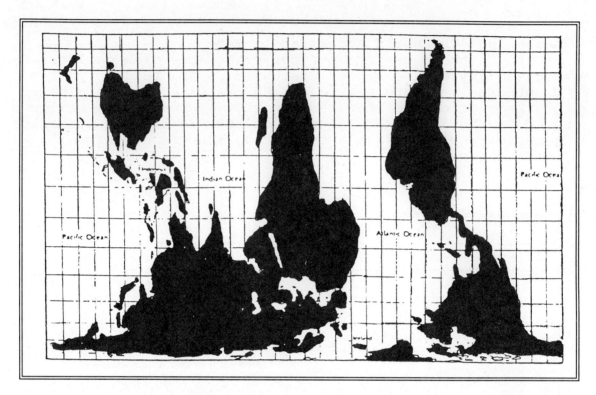

Figure 1-7 (repeated). Has this map been printed upside down? Or have Australians offered a corrective to our Northern bias? (From *A New View of the World*, by Ward Kaiser, copyright 1983 by Friendship Press, 475 Riverside Drive, New York, NY 10115.)

3. A black line drawing of the world (Figure 1-7) is reprinted above. It has obviously been repositioned from our normal view. According to friends in New Zealand, we have now drawn the world the way it was supposed to be seen all along.

   a. If you were charged with the job of coloring in the countries, how would you choose colors, or do you think there are better ways to color the map?

   Traditionally, the countries that were colonized by various 15th, 16th and 17th century powers had the colonies colored the same as the "parent" country. This is still true today even though many of the colonies are independent countries! Is this fair to them?

Answer to question #3a

b. What geographic signals are sent by the map orientation? How strong do you believe these are? Has a distorted view of the world hurt the progress of the science of oceanography by de-emphasizing and arbitrarily dividing the seas?

Answer to question #3b

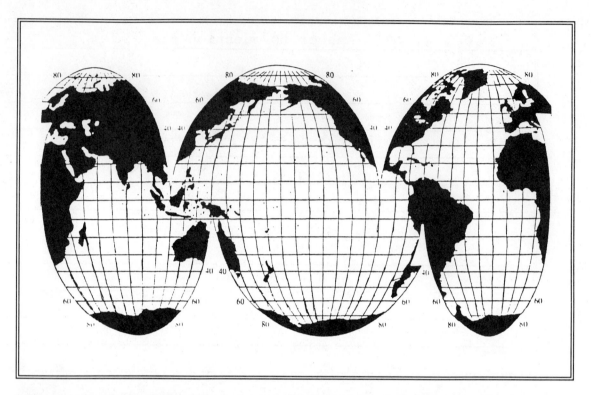

Figure 1-8 (repeated). A Homolosine World Map Projection. This map view implies that the world is mostly one solid ocean. (From *A New View of the World*, by Ward Kaiser, copyright 1983 by Friendship Press, 475 Riverside Drive, New York, NY 10115.)

4. Again, repositioning the world map to yet another perspective, this time so that the Pacific Ocean is highlighted, we see yet another view of the world (Figure 1-8) reprinted above. Do you believe your perspective of the world's ocean would be different had this view been in every classroom? Would people have a better respect for the seas?

Answers to question #4

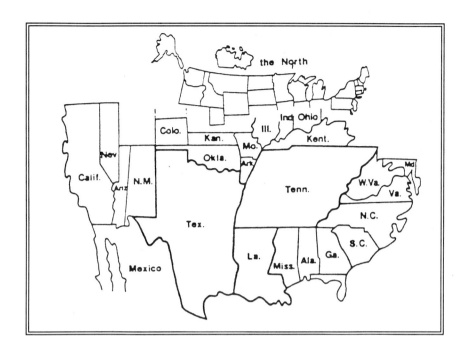
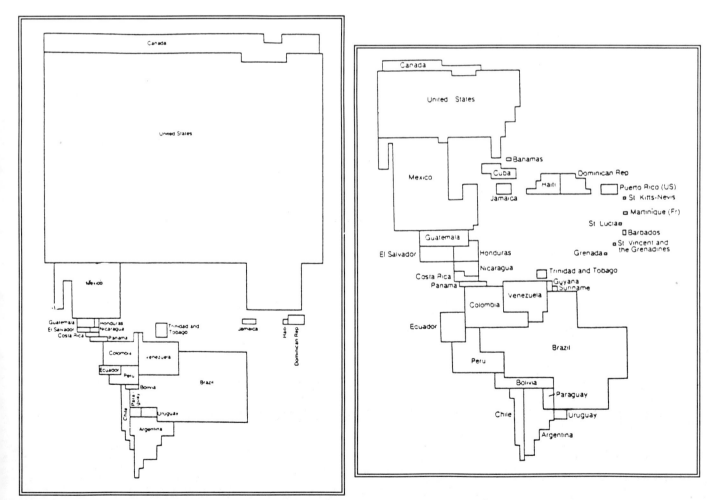

Figures 1-9 A, B, C.   Cartograms or quantified-information maps.
Citations follow Further Readings, pg. 21.

5. You have been shown several cartogram or quantified-information maps of the world (Figure 1-9 A, B, C). These are informational maps for purposes other than geography. They are drawn in a mock global manner to show global relationships. For what purpose do you believe that each was written? What does each show? Why or what clues do you have? Your instructor will provide the "correct" answers.

Answers to question #5
1-9 A.
1-9 B.
1-9 C.

Glossary of New Terms

Cartographer: A map maker.

Cartogram: Quantified-information maps presented in a pseudo world map image.

Further Readings

Campbell, J. *Map Use and Analysis.* Dubuque: Wm. C. Brown Publisher, 1991.

Dent, B.D. *Cartography Thematic Map Design.* Dubuque: Wm. C. Brown Publisher, 1990.

Fuller, B. *Fuller Projective - Transformation an interpretation accompanying the Dymaxion Map.*

Fuller, B. *Ideas and Integrities.* New York: Collier Books, 1969.

Kaiser, W. *A New View of the World.* New York: Friendship Press, 1983.

Thompson, M. M. *Maps for America.* United States Geological Survey, Third Edition, 1988.

Citation for Figure 1-9

A. By permission of *Harpers Magazine*.

B., C. By permission from The World Bank, 1984 publications.

*They that go down to the sea in ships,
That do buisness in great waters,
These see the works of the Lord,
And his wonders in the deep.*

Psalms 107:23-24

Laboratory TWO

Origin of Ocean Basins

Marine and Continental Rock Differentiation

To understand even the simplest reasons why the planet on which we live has surface-high places that are dry and low places covered by water, some basic information about the planet's formation and its early history is imperative. Information about some of the rocks that form the surface and their physical properties will be useful as well.

As the material that makes up the earth began to coalesce in space, the debris was of sufficient mass to have enough gravity to condense the planet into more or less its present size and shape. The compaction, due to gravity, formed heat throughout the planetesimal's interior - hot enough to melt some of the material in it. At the same time, meteorites and other material from space continued to bombard the surface, again releasing great amounts of heat and melting the material there. In general, heavier elements forming rocks sank to the center, while lighter rock forms rose to the surface. The earth formed into distinct internal bands that we refer to loosely as the core, mantle and crust. Each of these can be subdivided further. For our purposes, we shall only concentrate on the sub-divisions of the crust because this is where most of the action pertinent to oceanography takes place.

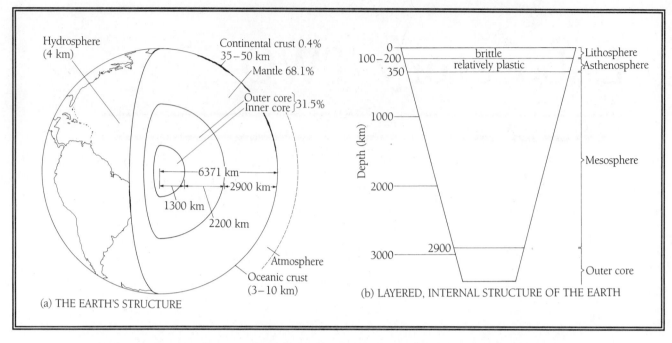

Figure 2-1 A and B. Subdivisions of the earth's interior. (from P. Pinet, *The Planet Oceanus*, West Educational Publishing, 1991.)

   The crust is a complex system of materials. It is layered yet has areas of intense vertical mixing not yet fully understood. Geologists feel, however, that there are recognizable differences at the outermost portion of the crust which greatly influence the landforms we see at the surface. These differences even control how we classify oceanic rocks from their continental counterparts. This differentiation will be discussed in greater detail later. For now, it is enough to say that there are differences in the materials that form large continents and ocean basins, in addition to differences in elevation.

   Figure 2-2 is a **hypsographic curve** for the earth. These curves show a distribution of various elevations between the highest and lowest points on the earth's surface. This difference is known as **relief**. On the earth there are over 13 kilometers of relief. A reference line, sea level, is included to differentiate those areas that are terrestrial from those that are marine. Continents cover about 35% of the earth's surface but nearly 20% of that amount is covered by ocean. The average elevation of the continents is 830 meters above sea level while the average ocean depth is much deeper, 3,700 meters below sea level. Still the hypsographic curve is remarkably bimodal; meaning it has two distinct areas of concentrated data points. We will see later how an apparently simple line, like sea level, actually causes some problems in interpreting relief of the world. Nearly everything physical in the world's seas ultimately came from the earth's interior; water, salt, and even the materials of the atmosphere.

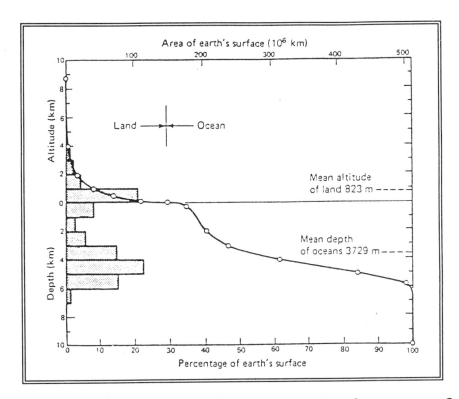

Figure 2-2. Hypsographic curve showing area of earth's solid surface above any given altitude or depth. Bar graph on left shows the distribution of area by 1-km intervals. (From Menard and Smith, 1966.)

It is not necessary to be a geologist to study physical oceanography. Geological processes are only part of the whole picture. An understanding of the chemical, biological and physical processes are important as well. Figure 2-3 shows in graphic from the interplay of disciplines involved in the study of oceanography. An introductory study of the materials comprising the earth's surface is necessary to the study of oceanography. Rocks and sediments made from rock degradation, or breakdown, from the perimeter of all our world's seas. Different rock types react differently to the same oceanic and atmospheric conditions. Therefore, if we know what the conditions are (waves, tides, salinity, prevailing wind direction, tectonic position, etc.) and something about the rock types involved, we can often predict the resulting land or seascape. For instance, rocky coasts composed of granite, such as those found along Maine, break down through erosion to form sand-sized particles similar to the original composition of the granite. Under the right conditions the particles collect and form beaches. However, this is not the only way sandy beaches can form. Under different conditions the same materials may not form beaches at all.

Rocks are made of aggregates of minerals. **Minerals** are inorganic crystalline solids of definite chemical composition found in nature. While this definition of minerals may be intimidating, it can be broken down to an easily understood group of terms.

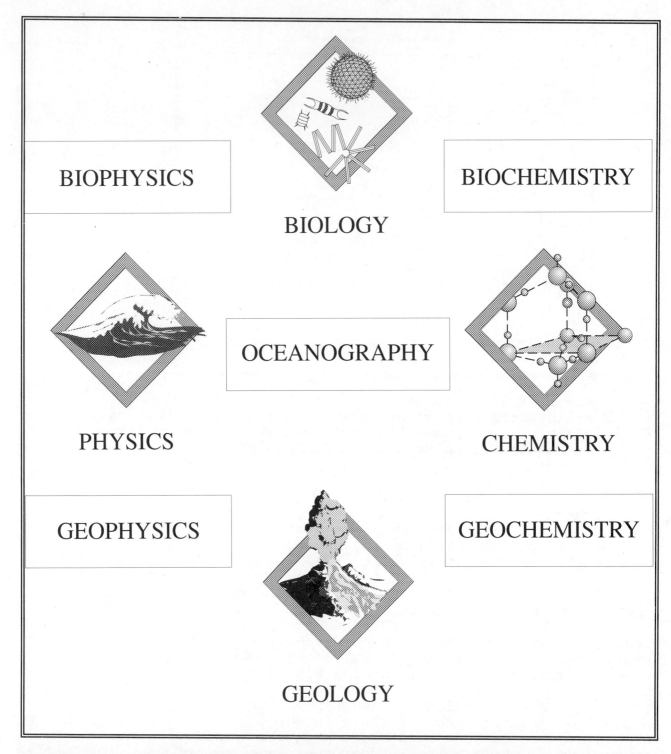

Figure 2-3. The fields of Oceanography (From P. Pinet, *Oceanography: An Introduction to the Planet Oceanus*, **West Publishing Company, 1992**).

Rocks are classified into three groups based upon how they formed. This may seem very straight forward at first, but there are problems in that rocks in some of the categories look and behave very much like rocks from other categories. In addition, the earth and all its components are in a constant state of change.

It is quite common for members of one group of rocks to be transformed from that category to another through the earth's natural processes. This is illustrated by the rock cycle diagram (Figure 2-4). Note that not only can rocks change groups, but they can also change identification within the same group by means of various physical and chemical activities. This too is shown by the rock cycle diagram (Figure 2-4).

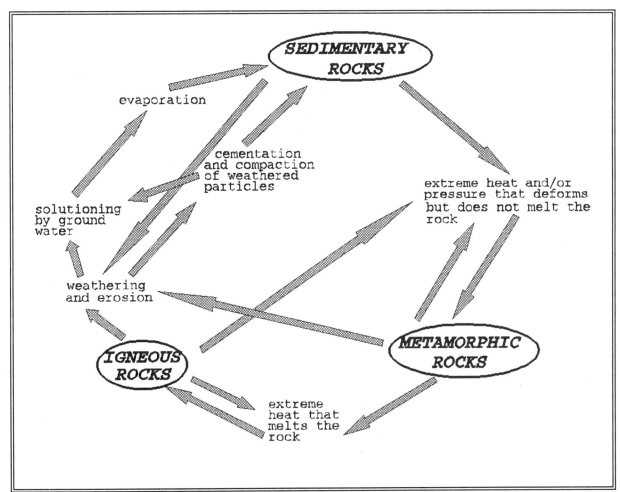

Figure 2-4. **The geologic rock cycle.**

The three basic rock types are **igneous, sedimentary,** and **metamorphic**. **Igneous** rocks are formed by the cooling of different types of magma. Magma is rock heated to its melting point, for most rocks this is over 1000º C. An example of cooled magma at the earth's surface would be volcanic lava. Much magma cools within the earth and never reaches the earth's surface. **Sedimentary** rocks are formed by either consolidation of the erosional products of surface rock (example: sandstone) or by precipitation from saturated solutions (example: rock salt). **Metamorphic** rocks are rocks that have changed in form by exposure to extreme heat (not to the point of melting), pressure and/or chemically active fluids. An example of this reaction is the alteration of limestone to marble.

As mentioned before, rocks are constantly being recycled to new positions in the rock cycle. The forces that alter the rocks to the points of change are many and include extreme heat, pressure, chemical activity, weathering and erosion. One can imagine where these forces might be found both on and in the earth. To have much of an effect, many of these forces must be applied over large amounts of time, even millions of years in some cases, while other forces (i.e. the effects of intense heat) can alter rocks instantly. The earth has had plenty of time since its origin 4.5 billion years ago for even the longest of these processes to occur.

The composition of igneous rocks can be said to come in two "types": those rich in quartz and feldspars (**Sialic** or **Felsic**) and others rich in quartz and dark colored minerals with high iron and magnesium content (**Mafic**). Felsic igneous rocks are created primarily from melted continental crust. When cooled, these rocks often assumed light colors and now make up the **rhyolite/granite** group. Mafic igneous rocks are composed mostly of oceanic crustal material, are often dark or black colored and are denser than felsic rocks. Mafic rocks make up the **basalt/gabbro** group. An intermediate group made up of mixtures of mafic and felsic rocks is called **andesite,** but they are much less common and are frequently difficult for introductory students to recognize. Andesitic rocks will not be addressed in this lab exercise.

Felsic igneous rocks are not only lighter colored, but, because they contain a smaller percentage of heavy metals, are also less dense than mafic igneous rocks. Scientists use the measurements of **specific gravity**, a comparison of a mass to an equal volume of pure water, to measure density. As both felsic and mafic rocks formed at the surface over the past 4.5 billion year history of earth, they separated into like types forming the rigid crust with lighter felsic rocks "floating" on the denser mafic mantle material. The felsic "islands" combined through random collisions and eventually, a few billion years ago, formed the continents. As mafic material cooled, it formed the ocean basins. Although both felsic and mafic material "float" on the denser mantle they are not equal in thickness. Continents average 35 Km (22 miles) in thickness but may be as thick as 60 Km (35 miles) below high mountain ranges. Oceanic crustal thickness is a bit more uniform and averages around 11 Km (7 miles) in thickness. The concept of crustal floating is called **isostasy** and can be compared to ice floating on water. You may remember that 90% of an iceberg is submerged, so too is the majority of the continents. About 55% of the continents displace mantle material. Perhaps a better analogy of the isostasy of continental material and its interaction with oceanic material is an ice cube resting on slightly thickened gelatin. Like the felsic and mafic material the ice and gelatin are composed mostly of the same ingredient (water for the food and quartz for the crustal matter). They are also different enough to be of different densities and have different properties. Figure 2-5.

To demonstrate this relationship mix up a box of your favorite flavor gelatin desert. As the gelatin begins to congeal place a large ice cube (a 1/2 lb butter tub works well if the gelatin is in a nine inch cake pan) centered on the gelatin. The ice and the gelatin will come to an equilibrium. The weight of the ice will depress the gelatin. The gelatin will then flow to areas in the pan with less pressure on it. Figure 2-6. This is very similar to the interaction of crustal materials. If conditions were sanitary during your demonstration, you now have dessert ready for the evening.

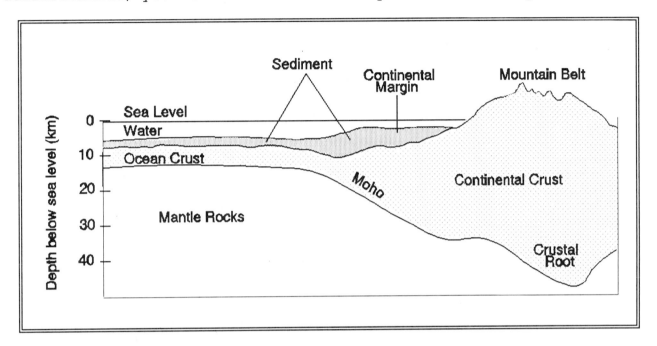

Figure 2-5. Idealized crustal profile showing isostacy in the earth's crust/mantle relationship.

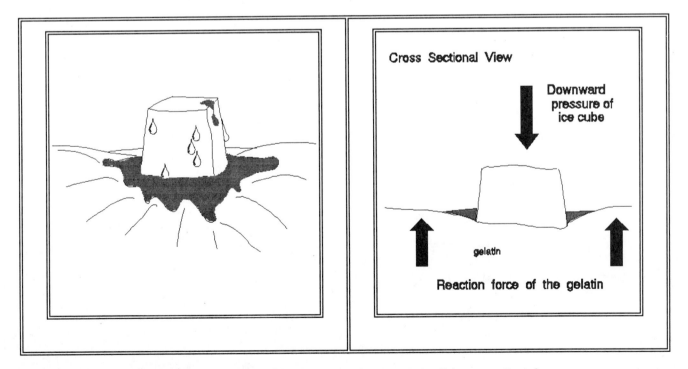

Figure 2-6. Isostasy of an ice cube in congealing gelatin.

Exercises

To this day, continents still have felsic igneous rock roots, while no such material is present in the ocean basins. You may find it interesting to note that even though Greenland is rather large for an island, it will always be classified as such because it is made up principally of the oceanic mafic rock.

1. **Determining specific gravity**
   Using the simple displaced water method (detailed below), determine the specific gravity for 1) a granite or rhyolite sample and 2) a basalt sample. For this you will need the following directions:

   a. Choose one sample of granite or rhyolite and one of basalt from the bins in lab. It may help you see the point of this exercise if they are nearly the same size. You will need two beakers or containers, one which easily nests inside the other. The rock samples must each fit into the smaller beaker and be completely submersible in water filling the beaker.

   b. Weigh the rock samples separately. Record your answers.

   ```
   granite/rhyolite sample weight _____
   basalt sample weight          _____
   ```

   c. Fill the smaller beaker to its brim with water and weigh the beaker and water. Record your answer. Weigh the plastic catch basin or larger beaker. Record your answer.

   ```
   water-filled beaker weight _____
   catch basin weight         _____
   ```

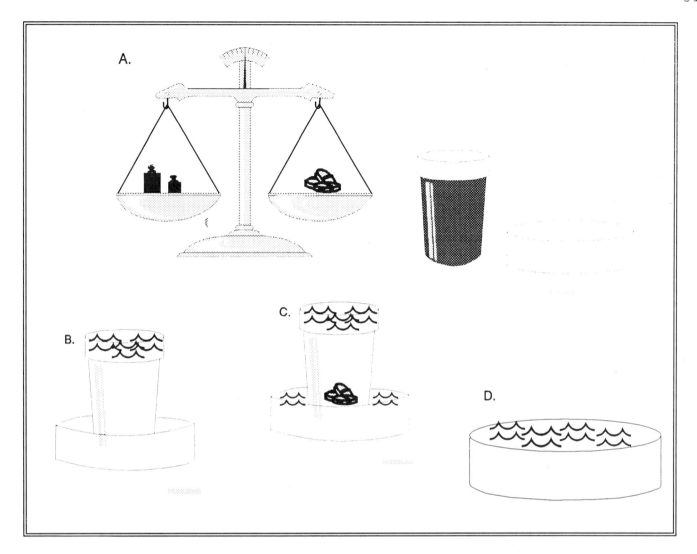

Figure 2-7. A. Weigh rock samples and each vessel used separately. B. Water-filled beaker in catch basin before sample added. C. After sample added. D. Weigh catch basin and spill-over water.

    d.   Place the water-filled beaker in the catch basin, carefully not allowing any water to spill. If it does spill go back to step C and begin again. After the water-filled beaker is in place, carefully lower (not drop) one rock sample into the water. The water in the beaker will spill out in to the catch basin. After the system settles down (a few seconds), remove the beaker with the rock & water carefully from the catch basin. Figure 2-7. Allow the water to drip off the beaker for a few seconds; then weigh the catch basin and water spill over. Record your answer. Repeat for other rock samples.

> catch basin and spill-over water weight _____

e. Complete the calculation below for each sample.

$$\frac{weight\ of\ rock\ sample}{(weight\ of\ catch\ basin + spillover\ water) - catch\ basin} = spec\ gravity$$

List each variable, and write up your procedures on a separate page. Take good notes during the lab demonstration so that you may follow proper procedure. Remember that specific gravity is a comparison of one object's mass to that of an equal volume of water. Therefore, by definition pure water has a specific gravity of one.

2. **Isostasy**
To see for yourself the property of isostasy, choose a scrap block of wood, blocks nearly cubic about 4 inches on edge work well. Weigh the block. Then float the block in a basin of water. Mark a water line directly on the block. Remove the block and wipe off excess water. Use a hand saw to make a cut along the drawn water line separating the block into two pieces. Weigh each piece and determine how much of the block was submerged using the formula below.

$$\frac{weight\ of\ submerged\ portion}{weight\ of\ entire\ block} \times .01 = \%\ submerged$$

Blocks of different woods (hardwoods vs. softwoods especially) will yield different percentages the same way rocks of different compositions will too. The equilibrium point of the wood sample and water is the percentage you determined.

Glossary of New Terms

Basalt/Gabbro Groups: Mafic igneous rocks that are dark colored. These make up continental crust.

Hypsographic curve: A graph showing a distribution of the elevation of the earth.

Igneous Rocks: Rocks formed from cooling of melted magma.

Isostacy: The concept of the crust floating like an iceberg on mantle material.

Mafic Rocks: Igneous rocks rich in iron, magnesium and quartz that form oceanic crustal material.

Metamorphic Rocks: Rocks formed by alteration of other rocks by heat (below melting point), pressure and chemical activity over time.

Mineral: An inorganic crystalline solid of definite chemical composition found in nature.

Relief: The difference between the highest and lowest elevation in an area.

Rock: An aggregate of minerals.

Ryolite/Granite Group: Sialic (felsic) igneous rocks that are usually light in color. These make up continental crust.

Sedimentary Rocks: Rocks formed from fragments of other rocks or from precipitaion from a solution, usually water.

Sialic (Felsic) Rocks: Igneous rocks rich in quartz and feldspars that form continental crustal material.

Specific Gravity: A comparison of density measured against an equal volume of water.

Further Readings

Ballard, R.D. *Exploring Our Living Planet*. National Geographic Society, Washington, D.C., 1983.

Barghoorn, E.S. The oldest fossils. *Scientific American.* 224:5, 30-54, 1971.

Bullard, Sir Edward. The origin of the oceans. *Scientific American* 221:66-75, 1969.

Hamblin, W.K., and Howard, J.D. *Physical Geology Lab Manual*. Edina, MN: Burgess Publishing, 1986.

Hess, H.H. *History of ocean basins. Petrologic studies: A volume to honor A.F. Buddington*, Engle, A.E.J. York: Geological Society of America, 1962.

Menard, H.W. The deep ocean floor. *Scientific American.* 221:53-63, 1969.

Menard, H.W. and S.M. Smith. *Hypsometry of Ocean Basin Provinas*. Journal of Geophysical Research. 71:4305-4325.

Pinet, P. *Oceanography: An Introduction to the Planet Oceanus*, West Publishing Company, 1992

Schlater, J.G., Parsons, B., and Jaupart, C. Oceans and continents: Similarities and differences in the mechanisms of heat loss. *Journal of Geophysical Research.* 86:11, 535-52, 1981.

Press, F., and Siever, R. *The Earth*. San Francisco, CA: W.H. Freeman and Company. 1974.

Tennison, J. A Chronicle of Carbonate Platform Development. *Geotimes*, 36:4, 1991.

Laboratory THREE

Marine Sediments

And Rocks Formed from Them

Over three-fourths of the materials covering the earth's surface and an even higher percentage of the sea floor are sediments and the rocks formed from sediments. Sedimentary rocks occur in two general forms, **clastic** and **nonclastic**. **Clastic** sedimentary rocks are formed from weathered and eroded materials. Rocks made from particles transported only a short distance from their source tend to be larger and have sharper edges than those transported a longer distance. Those clastic rocks that have been exposed for only short periods of time will have greater relative amounts of easily weathered materials associated with them. In sialic rocks, those that make up continents, quartz is the mineral most resistant to weathering. If this is true, one might expect to find massive accumulations of nearly pure quartz particles collected at the mouths of rivers that traverse areas high in granites and on the beaches where those rivers deposit their loads. This is exactly what we see when we go to many beaches. We will see examples of this kind of beach sand in a later lab exercise.

Sediments and sedimentary rocks are classified using three main criteria:

- Size of grains. Since most particles have been transported by wind or water, they have also been **sorted** to uniform sizes by the changing velocity of whatever is moving them. This allows us to use grain size not only as a tool in classifying these rocks but also in determining what source of energy deposited them; i.e. winds seldom move boulders. Figure 3-1.

- Roundness (sphericity or angularity) of the grains.

Because of the difficulty in seeing this feature in very small particles, roundness can only be recognized in larger particles and is used for those grains of sand-size or larger (2 mm, the size of a pin head or larger). The terms used to describe sphericity are a continuum from very angular, angular, sub-angular, sub-rounded, and rounded, to well-rounded.

o Composition of the grains, or the ratio of minerals that make up the particles. The terms used to describe composition most frequently are: mostly quartz, mostly non-quartz rock fragments or a mixture of rock fragments, and quartz grains.

Another less important criteria in naming the sediment, but very important in determining its origin, is mixture. This refers to the uniformity of size, whether the grains are unimodal, bimodal or trimodal in distribution.

SIZE CLASSIFICATION OF SEDIMENTARY ROCK GRAINS			
Name of Particle	Diameter Range	Rock Type	
Boulders Cobbles Pebbles Granules	greater than 256 mm (±-10 in.) 64 - 256 mm (±-2.5 - 10 in.) 4 - 64 mm (±-0.15 - 2.5 in.) 2-4 mm	<u>Rounded</u> *Conglomerate*	<u>Angular</u> *Breccia*
Very coarse sand Coarse sand Medium sand Fine sand Very fine sand	1 - 2 mm individually ½ - 1 mm recognizable to ¼ - ½ mm naked eye ⅛ - ¼ mm 1/16 - ⅛ mm	*Very Coarse* *Coarse* *Medium* *Fine* *Very Fine*	*Sandstone*
Coarse silt Fine silt (gritty feel; grains cannot be seen)	1/64 - 1/16 mm 1/256 - 1/64	*Siltstone*	
Clay (smooth feel)	less than 1/256 mm less than 1/256 mm	*Shale*	

Figure 3-1. Sedimentary rock grain classification by size.

Nonclastic sedimentary rocks are formed through chemical and/or organic processes. A nonclastic sedimentary rock occurs when ions in solution (usually water) become super saturated and precipitate out in solid form. This is often due to evaporation or changes in the temperature of the solution. Rock salt and limestones, common marine rocks, are examples of non-clastic sedimentary rocks. Some limestones are also formed by biological processes. These can be most easily seen in areas where organic reefs develop, as they have along Andros Island in the Bahamas and the in Great Barrier Reef in Australia. The biologically produced limestone comes primarily from calcareous algae, shells of mollusks and the internal skeletal material of corals. The biotic life forms precipitate the calcareous particles organically, directly from the calcium rich water. This process is in contrast to inorganic precipitation resulting from evaporation and saturation. Nearly always the biologically produced reefal limestone can be recognized by the fossils it contains.

Seeing the relationship between marine deposits like barrier islands or beaches and the original terrestrial material that make up these deposits is difficult. To assist seeing those relationships a short narrative is provided below.

JOURNEY OF AN IGNEOUS ROCK TO A BEACH

High in the Rocky Mountains a large granitic boulder breaks from the side of a steep incline and falls near a stream bed. Through seasons of freezing and thawing this boulder is mechanically and chemically weathered and begins to crack and break apart. Some of the smaller pieces of rock are carried into the stream by rivulets of rain water, but many pieces are too heavy to be transported rapidly.

The stream cuts away sand behind the larger pieces, causing the rock to roll and travel slowly down the stream bed. All the while this granitic rock is being further eroded by the abrasion from the sand-sized particles in the stream's column, the motion of the water, and the chemical agents within the water. Of course, the less resistant materials are the first to be broken down. The rock is broken into still smaller parts as it travels down tributaries to the main stream which eventually empties into the ocean. By the time the rock reaches this point only the most resistive materials are left, quartz and a few flakes of mica in the case of granite.

Exercises

1. **Determining Sediment History**
   There really is a difference in the distribution of particles in various sedimentary environments. In the early 1950's Robert Folk, University of Texas, recognized that if one could determine the distribution of the various sized particles in sediments by percent weight, the results could be plotted graphically such that when compared to similar plots of material from known environments, the deposit's history could be determined. This recognition was based on three observations.

   ○   Different modes of transport move and deposit materials differently.

   ○   There are a variety of different environmental settings where sediments are deposited.

   ○   As long as a wide range of material sizes are available, then the patterns formed by the plotting of the data in some forms were nearly unique to specific environments. Beaches, rivers, sand dunes and many other sedimentary environments have processes unique to each that can be seen on a graph of the sediments by size and percent weight.

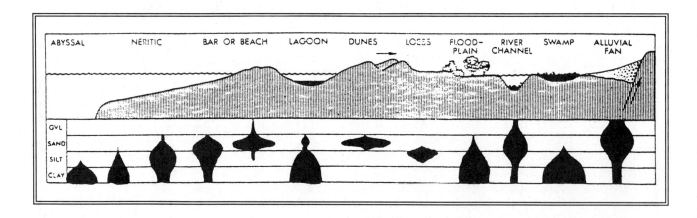

Figure 3-2. Spindle diagrams for size distribution of sediments in twelve sedimentary environments of deposition. Author unknown.

Figure 3-2 shows a gross distribution of twelve sedimentary environments in spindle diagram plots. Spindle diagrams are so called because they look like the turned wood pieces or decorative spindles on stairways and furniture. The spindle is made by plotting data along a centered vertical baseline, usually zero. The eight environments on the left side of the figure are all related directly or indirectly to marine activities.

Answer the following questions based on your observations of the chart:

a. What is the general trend of sediment size distribution from the abyssal planes to the beach? What causes this trend?

Answer to Question #1a.

b. This graph only shows relationships of size. Do you believe there are other parameters that, when plotted, could show different trends? Consider roundness and degree of sorting (dominance of one or two size ranges)? What might those trends look like or show over such a range of environments as those shown in figure 3-2.

Answer to Question #1b.

c. What might cause well-rounded grains to appear in a poorly sorted sediment? Poorly sorted sediments are those with a wide variety of sized sediments contained within them. They often indicate deposition near the source of the parent rock. Well rounded grains usually indicate long term transport.

Answers to Question #1c

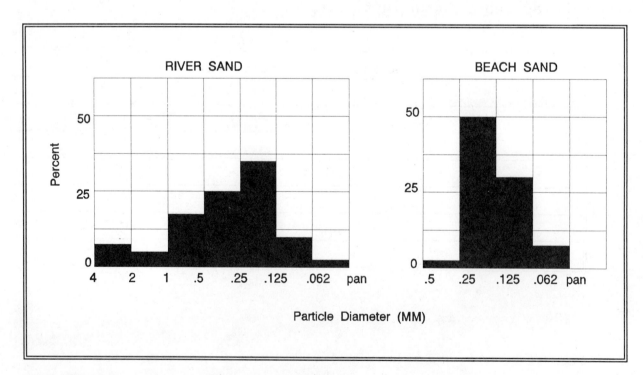

Figure 3-3. Grain size distributions of beach and river sediments.

2. **Recognizing Limestone**
The most common non-clastic marine sedimentary rock found in the rock record is limestone. Fortunately there is a very easy and diagnostic test for its presence. Limestone is made from the mineral calcite. Calcite reacts by effervescing (bubbling) in the presence of Hydrochloric Acid (HCl). For this test weak concentrations of HCl work as well as strong ones and are less dangerous. The fumes of even the weak acid concentrations can be unpleasant and harmful. Be careful to use a fume hood or at least a well-ventilated area when performing this exercise.

Using a weak (2-3% concentration of technical grade, 37% assay HCl) solution of hydrochloric acid, place a few drops on the samples of limestone provided. See the reaction. Rinse the rock and resultant solution down the drain with plenty of water after each application of acid. This concentration of solution will not burn skin unless it stays in prolonged contact. Wash area thoroughly in water if skin comes in contact. The most frequent accident using the acid is more annoying than dangerous. Acid dropped on clothes will bleach the dye out of the cloth. Flush any drops on clothes quickly with water.

A variety of different types of limestones will be seen in later lab exercises. Sometimes the mineral calcite will be the cement that holds sand grains together in sandstone. These rocks are frequently confused with limestones because the calcite cemented sandstones will also fizz in the presence of an HCL solution. To tell the sandstone from limestone see if there are grains of rock fragments present. If so, the sample is probably a sandstone. Apply the acid solution and see if it is the **grains** that are effervescing, if not the sample is most likely a sandstone. A magnifying glass may be helpful.

3. **Comparing River and Beach Sand.**
Figure 3-3 shows typical distribution plots for sand from a river and from a beach. You will be provided with a sample of unknown sand to sieve through a series of mesh screens with holes of known size openings. These screens or sieves correspond to the sizes shown on the graph plots; 4, 2, 1, 0.5, 0.25, 0.125, 0.062mm.

Procedures

○ Weigh the entire sand sample and container.

> Sand and container weight _____

○ Sieve the sand in a Ro-Tap (a mechanical sieve shaker) for 15 minutes (manual sieving will work but accuracy may suffer slightly). Fifteen minutes is considered a standard constant and was determined by Folk experimentally to be the optimum time duration.

○ Weigh the empty container that held the sand sample to determine the sand's weight. Use the following formula:

Sand + container weight - container wt = wt sand

Empty container weight_____

○ Weigh the sand from each sieved size fraction, (be certain to subtract container weight).

Sand weight 4.0 mm sieve
Sand weight 2.0 mm sieve
Sand weight 1.0 mm sieve
Sand weight 0.5 mm sieve
Sand weight 0.25 mm sieve
Sand weight 0.125 mm sieve
Sand weight 0.062 mm sieve

○ Plot, on the graph provided, the results in bar graph form.

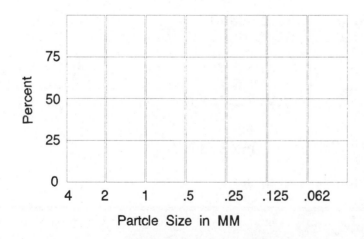

○ Compare your results with the graphs on figure 3-3 and determine probable source for this sample.

Obviously, the results you generate may not be identical to either of those in figure 3-3, but they should more closely resemble one than the other. Answer the following questions about your experiment.

a. Why might the results from your unknown sand plot differ from those in figure 3-3?

Answer to Question #3a

b. Is a terrestrial source like a river the only source for beach sand? Explain your answer.

Answer to question #3b

c. What errors in lab procedures could cause errors in interpretation?

Answer to question #3c

Determining the history of sediments is not a trivial matter. It supplies scientists with information about the tectonic history of the area and clues to possible mineral, fuel and other valuable resources which may be contained within the sediments or in nearby deposits.

4. **Sediment Stratigraphy/Graded Beds**
    Several sealed jars of water and sediments of various sizes, shapes and colors have been prepared for you.

    a. Agitate a jar of sediment and let it sit for a few minutes. Notice the layers developing. How may layers form? Draw what you see in detail and try to explain your observations.

    Drawing of layers in jar. Label layers to the right of diagram.

    b. Next slowly tilt the jars until the sediment layers begin to flow. Is the flow even and uniform? Why or why not? Of what importance is this observation to the study of sea floor sediments?

Answer to question #4b

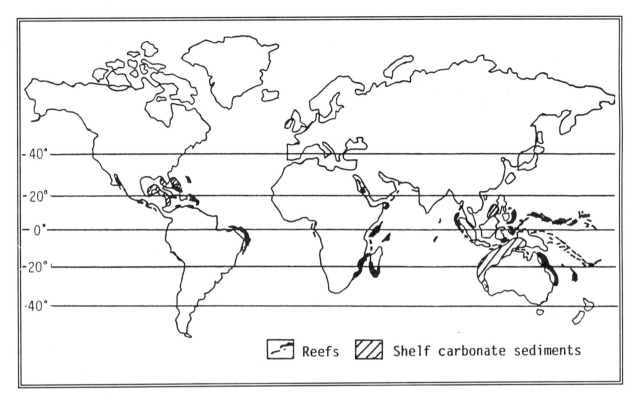

Figure 3-4. Major reefs of the world.

5. **Barrier Reef Distribution**
Your lab text discusses non-clastic production of limestone by biological means. The map provided (Figure 3-4) shows where the major reefs on the globe are located. What conditions do you think affect the production of reefs (e.g. warm or cool temperatures)? What other factors might be important?
List at least five and discuss them. Don't run to your text here; this is meant as a question to ponder. This question will be discussed at length later in the course, as this is an area of critical concern.

Answer to Exercise #5

Glossary of New Terms

Clastic Sedimentary Rock: Sedimentary rocks formed from the particles weathered and eroded rocks. Examples: conglomerate, sandstone and shale.

Nonclastic Sedimentary Rock: Sedimentary rocks formed by biological or chemical precipitation. Examples: limestone, dolostone and rock salt.

Further Readings

Folk, R.L. *Petrology of Sedimentary Rocks*. Austin, TX: Hemphill Publishing Co., 1980.

Kennett, J.P. *Marine Geology.* Englewood Cliffs, New Jersey: Prentice-Hall, 1982.

Pinet, P. *Oceanography: An Introduction to the Planet Oceanus*, West Publishing Company, 1992

Pettijohn, F.J. *Sedimentary Rocks*, 3d ed. New York: Harper & Row, 1975.

Shepard, F.P. *Geological Oceanography: Evolution of Coasts, Continental Margins and the Deep Seas Floor.* New York: Carne, Russak, 1977.

Siever, R. *Sand*. New York: Scientific American Library, 1988.

Laboratory FOUR

Plate Tectonics

Plate Tectonics and Coastal Development

The single most important advance in the study of coastlines in the past several decades has been the discovery that the surface of the earth is in constant motion and that indeed oceans are and have been expanding and shrinking since the time of their formation over two billion years ago. The concept that continents have moved with respect to one another over time is not entirely new. Anyone who has carefully viewed the earliest maps showing the western coast of Africa and the eastern coast of South America may have felt that the similarities in their shape was more than coincidental. Surely, some time in the past, these two land masses must have been conjoined. However, this idea did not appear in print until Antonio Snider-Pellegrini suggested it in his book, *La Cre'ation et Ses Myste'res Devoil'es*, in 1858, that all the continents may have formed into a single landmass during the late Paleozoic Era and have since split apart. Figure 4-1. He believed that the rupturing of the continent occurred as a result of the Noahan Flood recounted in the book of Genesis in the *Bible*.

Several wrote on the idea that continents have moved through time in the intervening years. However, it was not until 1915 when Alfred Wegener published his book, *The Origin of Continents and Oceans*, that the **Theory of Continental Drift** was truly developed. He coined the term, **Pangea,** used for the supercontinent of all the lands together that Snider-Pellegrini envisioned. Wegener was a meteorologist, and he used information based on climatology, paleontology and geology to support his theory. Although Wegener had many detractors, he had a few ardent supporters including the famous South African geologist Alexander du Toit, who provided additional evidence to support Wegener's claims. Most geologists refused to believe Wegener's theory and felt there was not enough energy on or in the earth to facilitate continents in plowing through the dense mafic rocks of the ocean floor. The nonbelievers were partially correct, for in the 1950's, the discovery was made that the continents did not plow through the ocean floors as they moved. Rather, they were broken and pushed apart as new rock was formed along rift systems in the center of some of the land masses. Then, the continents were pulled along by the forces subducting older material along the plate margins.

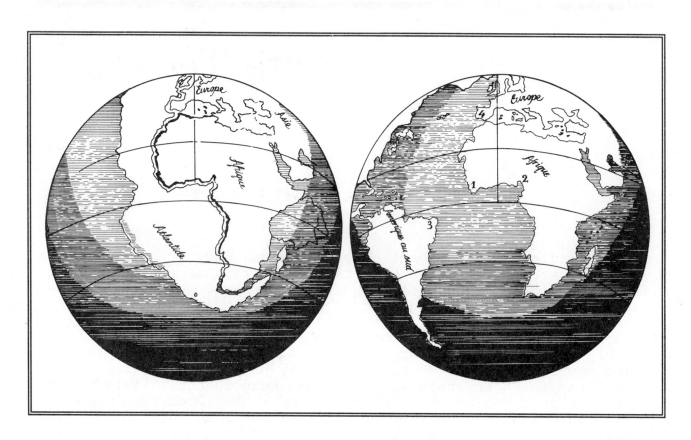

Figure 4-1. Reproductions of maps published by Antonio Snider-Pellegrini in 1858. (From P. Pinet. *Oceanography: An Introduction to the Planet Oceanus*, **West Publishing Company, 1992.)**

The mechanisms involved in the formation and destruction of crustal material is still something of a mystery, but more information is being added daily. This new theory, known as the **Theory of Plate Tectonics**, involves a series of rigid plates that move independently of one another. The general belief is that only 20 or so major plates are on the earth. The borders of the plates were determined primarily by analysis of earthquake epicenters or points of origin (Figure 4-3). Because the areas along the rigid plates are constantly trying to reach some form of equilibrium, constant motion results in massive earth movement along their boundaries. Some of the plates have continents riding on them, and some, like the Pacific Plate, have only ocean material on them.

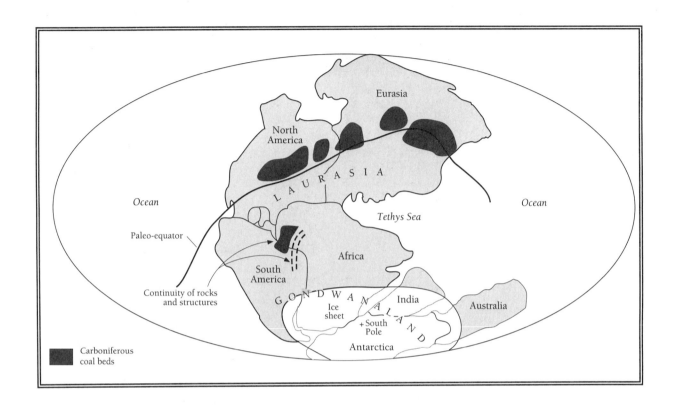

Figure 4-2. Alfred Wegener's 1922 reconstruction of Pangea using correlation of Early Mesozoic glacial deposits and coal beds. (From P. Pinet. *Oceanography: An Introduction to the Planet Oceanus* **West Publishing Company, 1992.)**

A thorough understanding of plate tectonics is not imperative for introductory students in oceanography, yet you should understand the basic implications of this revelation. Since its discovery and subsequent acceptance across the vast majority of the scientific community, plate tectonics has become a new **paradigm** to which all new ideas must either defer or disprove. It has provided a whole new perspective of the world for scientists and particularly physical oceanographers.

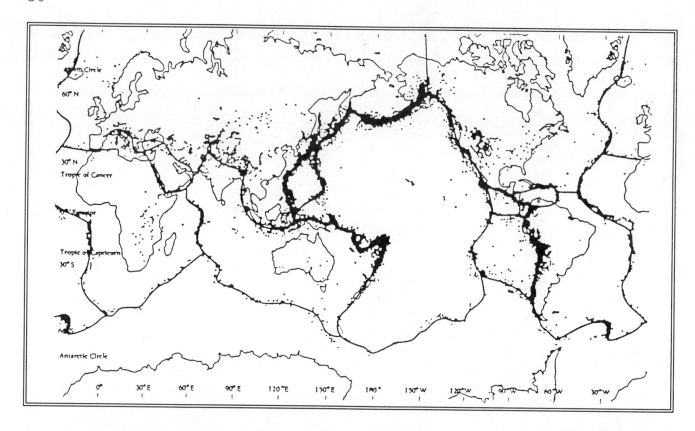

Figure 4-3. A world distribution of earthquakes from 1961-1967. Maps of the 70's and 80's show the same pattern. (Adapted from B. Isacks, J. Oliver and L.R. Sykes. *Journal of Geophysical Research.* 73, 1968.)

As plates jostle about for millions of years some inevitably will collide. As the plates bang against one another, one of three events will occur:

1) The plates will deflect off each other and consequently collide with an adjacent plate.

2) The energy driving the plates will become depleted, and the plates will remain in temporary quiescence.

3) The plates will continue to press against each other until one finally overrides and/or deforms the other.

When the third possibility occurs, crustal material is destroyed and reabsorbed into the earth's mantle. When this happens to plates with continents riding on them, the plate with the greatest density subducts below the less dense plate (Figure 4-4). We discussed the density difference between oceanic (mafic) and continental (sialic) material earlier in the lab on ocean origins. The continental material is much more difficult to subduct; but the effects of subduction can clearly be seen on coastlines of continents involved in such collisions.

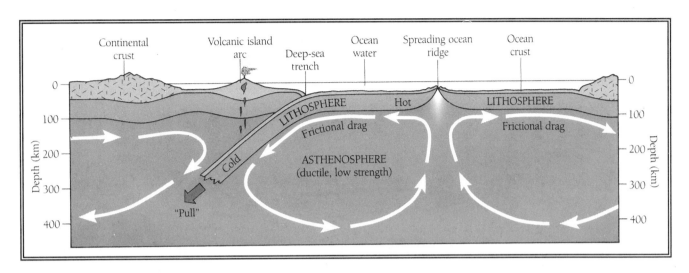

Figure 4-4. Schematic drawing showing sea-floor spreading. (From P. Pinet. *Oceanography: An Introduction to the Planet Oceanus*, West Publishing Company, 1992.)

As continents ride along with the plates, the leading edge or **active margin** is affected quite differently than the trailing or **passive margin**. Frequently, the active margin has coastal mountain ranges and a deep trench directly offshore where the subduction is occurring. There is almost no coastal plain, continental shelf, or slope present. The passive margin normally has extensive coastal plains and long, low-angled continental shelves and slopes.

Exercises

1. **Analysis of Plate Movement**
   The outline map of the world provided for you (Figure 4-5) has the major plate boundaries delineated for you.

   a. Study the features on the physiographic map of the world (National Geographic Society or others). Determine the direction each of the major plates is currently moving by analyzing features common to active or passive margins. Draw arrows directly on the map. For the plates marked A-E, justify your choice of direction in the spaces provided below. You may wish to work in pairs and even consult other pairs. They may offer different perspectives than your own.

   Hints to look for that may help in this endeavor are:

   - Location of active volcanos
   - Mountain chains parallel to and along coasts
   - Position of wide, flat coastal plains

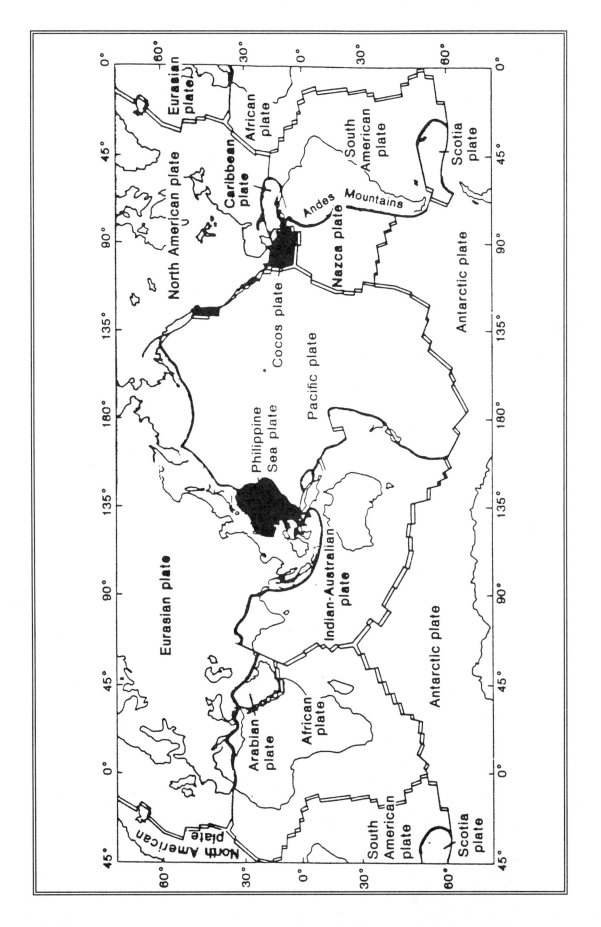

Figure 4-5. Map of the world with plate boundaries delineated. (From Charles C. Plummer and David McGeary, *Physical Geology*, 5th ed. Copyright(c) 1991. Wm C. Brown Publishers, Dubuque, Iowa. All rights reserved. Reprinted with permission.)

Direction of the Plates

A. Pacific Plate:

B. South American Plate:

C. Arabian Plate:

D. African Plate:

E. North American Plate:

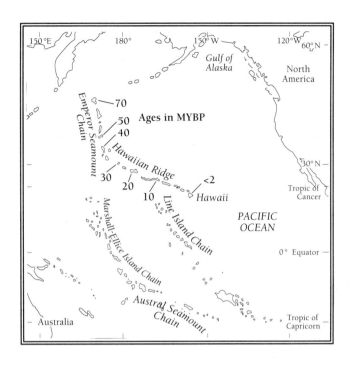

Figure 4-6. The Hawaiian Island Chain and ages for the island features. (From P. Pinet. *Oceanography: An Introduction to the Planet Oceanus,* **West Publishing Company, 1992.)**

b. Now look carefully at the Hawaiian Islands. Are there any clues that indicate the direction of the Pacific Plate?

Answers to exercise #1b

c. How does the ridge formed at the center of the Atlantic Ocean aid in determining the direction of movement along its margin?

Answer to exercise #1c

2. **Analyzing Plate Tectonic Map**

   a. From your observations above determine which ocean basins are shrinking and which are growing.

   Answer to exercise #2a

   b. The oldest oceanic rock found off the eastern coast of the United States and the oldest oceanic rock off the North Western coast of Africa are both Late Triassic in age (approximately 200 million years ago). Find the distance between coastal North Carolina and Western Sahara (Southern Morocco).

   Distance in miles from North Carolina to Western Sahara _____

   Assuming a uniform rate of spreading from the Mid-Atlantic Ridge over the duration since late Triassic time, how fast is the Atlantic Ocean enlarging itself? Remember there is growth on each side of the ridge simultaneously.

Answer to exercise #2b

3. **Optional Lab Exercise**
   Now that you think you have a sense of where the plates are going, we shall see how accurately you have determined the plates' current paths and see what course they have taken to get to their present positions. We will be using a computer simulation of the plates' movement throughout the last 500 million year history of the world. The program we have chosen is called **Time-Machine Earth** and is offered through Sageware Corp. It will allow you to choose any time during the last 500 million years and any position on earth to use as a vantage point. Follow the directions provided with the software package; it will answer most of your technical questions. For the most part you are encouraged to try any time/place coordinates you wish; however, you will be asked to make a few observations on coordinates provided.

A. S 05 latitude   Time Begin: 100,000,000 BC
E 90 longitude   Time End: Present
(Grid + Over commands ON) Travel Increment: 20 mill year
What great event are you watching form?
Is it still going on?

B. S 90 Latitude E 90 Longitude (Grid + Over commands ON)	Time Begin: 80,000,000 BC Time End: Present Travel Increment: 20 mill year

What 3 continents are the focus of this sequence run?

What happened to their relative positions during the last 80 million years?

Does observing this sequence give you any clues about why some continents have an assemblage of animals on them that are unique? (Hint: Keep in mind that marsupial mammals are believed to originate in South America but are seldom there today.)

This next sequence will concentrate on the western coast of the United States. Watch the development of the San Andreas Fault zone. Note that the continent of North America begins in this sequence moving west. But, as the program speculates up to 60 million years into the future, North America begins to move eastward causing a shift in the active and passive margins.

The program relies heavily on the concept of *status quo*. Rates and directions of plate movement remain constant as the program speculates the future. Notice that Southern California **does not** fall into the sea, but moves along the fault line to the north. The sequence graphics get confusing on this exercise when run in the proper time order. Try reversing the order by beginning at 60 million years in the future and working back to 40 million years ago. If you keep the over print on during this run, it may help you.

C. N 30o Latitude   Time Begin: 40,000,000 BC
    W 90o Longitude   Time End: 60,000,000 AD
    (Grid + Over commands ON) Travel Increment: 20 mill year

Glossary of New Terms

Continental Drift: Theory preceding plate tectonics that suggested that continents had changed their relative positions throughout geological time.

Pangea: "Supercontinent" of Alfred Wegener who envisioned all continental material conjoined at some time in the geological past.

Plate Tectonics: Theory used to explain movement of continents having several rigid plates moving independently on mantle material.

Paradigm: The principle model used to explain observations.

Active Plate Margin: Leading edge of crustal plate.

Passive Plate Margin: Trailing edge of crustal plate.

Further Readings

Allegre, C. *The Behavior of the Earth.* Cambridge, Mass.: Harvard University Press, 1988.

Dietz, R.S. and Holden, J.C. The breakup of Pangaea. *Scientific American* 223:30-41, 1970.

Dietz, R.S. Those shifty continents. *Sea Frontier.* 17:4, 204-12, 1971.

Mark, K. Ocean fossils on land. *Sea Frontier.* 18:2, 95-106, 1972.

Marvin, U. *Continental Drift, the Evolution of a Concept.* Washington, D.C.: Smithsonian Institution, 1973.

Pinet, P. *Oceanography: An Introduction to the Planet Oceanus,* West Publishing Company, 1992

Sullivan, W. *Continents in Motion. The Earth Debate.* New York: McGraw-Hill, 1974.

Wegener, A. *The Origin of the Continents and Oceans* (trans. from the 3d German ed. by J.G.A. Skerl). London: Methuen & Co., 1924.

Wegener, A. *The Origin of Continents and Oceans.* New York: Dover Publications (paperback reprint of original translation from 1915 original), 1966.

Laboratory FIVE

Bathymetry

Topographic Maps and Bathymetric Charts

In an earlier laboratory we discussed **relief,** as the term is used by geomorphologists (geologists who study landforms) and physical oceanographers. Remember that relief is the difference in elevation between the highest and lowest positions within a prescribed geographic area. If you look outside your window, you will probably see at least a few places that are higher or lower than the rest of your line of sight. That is the relief of the area. The information that a given area is not flat is often of great interest. For example, if you were considering buying property on which to build a house, you may be limited as to the style you could build by the roll of the landscape. Or, if you were planning a vacation of hiking in the mountains, it might be of interest how high you may have to climb on your journey. You need not visit each area that interests you to find this out. Fortunately, there are maps which address and provide this information for you. The United States Geological Survey, a division of the Department of the Interior, has and is preparing **topographical** maps of the entire United States.

Topographic maps use **contour lines**, continuous lines connecting points of equal elevations, to represent a three-dimensional surface on a two-dimensional map. Contour lines can be thought of as closed geometric figures, like circles and squares, that outline planes of equal elevations. One rather whimsical way to demonstrate contour lines might be to imagine a multi-layered birthday cake in the shape a comic strip bear. If you were to make such a cake yet only had a supply of one inch cake pans, you would have to build up various portions of your cake one equal layer at a time. To minimize baking too many cake parts you could design a plan for each layer that might look like Figure 5-1. A contour map of the completed cake then would look like figure 5-2. Figure 5-3 shows a cross-section of the cake.

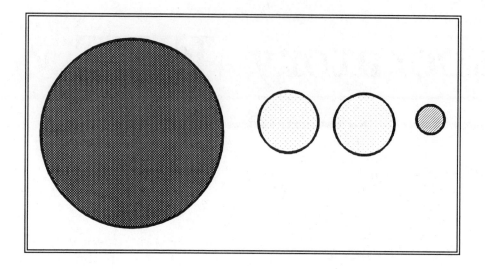

Figure 5-1. Pattern for a three layer birthday cake bear.

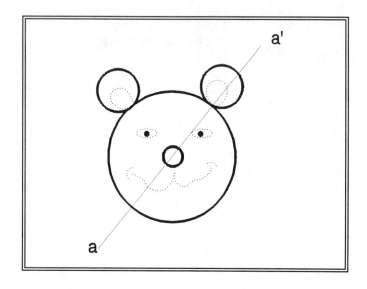

Figure 5-2. Final birthday cake bear with frosting highlights in dashed lines.

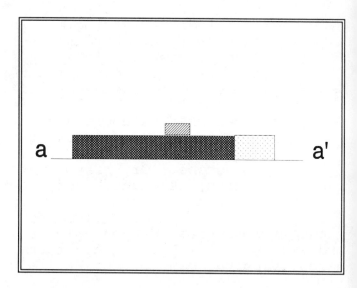

Figure 5-3. Cross sectional view of birthday cake along line a-a'.

An example of natural contouring might be seen along a lake that is being drained out like a bathtub. You will do a similar exercise during this lab period. Difficulties in visualizing this example will become clearer at that time. If the lake begins at a full level with a shoreline denoted by a beach, we can consider the beach as the contour line of the basin at level X. If one meter is drained and a new beach forms, that beach now becomes the contour line for X minus 1 meter. If we continued draining the lake in one meter intervals, allowing beaches to form between each draining, several beach/contour lines would be formed. If we took a photograph of the now drained lake from the air, we would see a series of concentric closed geometric forms delineated by the stranded beaches. If any portion of the lake had steeper margins than other areas of the lake, these would show up on our photograph as beaches/contour lines much more closely spaced than those areas with shallow grades or inclines. Thus, we can see that the more closely the spacing between contour lines, the steeper the grade (Figure 5-4).

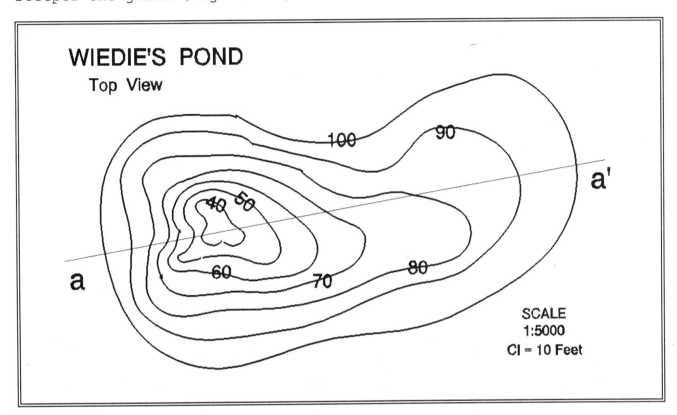

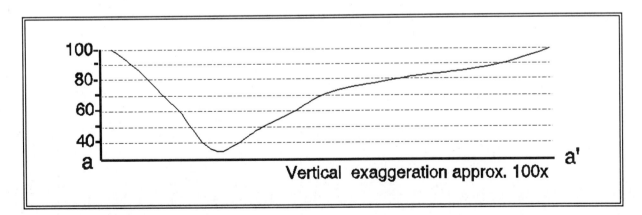

Figure 5-4. A. Top view of stranded beaches in drained lake. The feeder stream in all likelihood extends to the top right of the figure as currently oriented. B. Cross sectional view along transverse A-A'.

One can also see on our aerial photograph that the beaches never split or cross one another. This is true for contour lines as well. Additionally, you can tell by the contour lines/beaches where the feeding river is for the lake we just photographed. As a rule when contour lines cross linear depressions, like valleys or streams, the contour lines themselves will always point up-slope or up-stream. This axiom is frequently referred to as the **rule of V's**.

Contouring Rules

○   Contour lines show and connect points of equal elevation.

○   Contour lines never bifurcate or split.

○   Contour lines never cross one another.

○   Contour lines V or point up-slope or up-stream when encountering valleys.

In addition to the contour maps made of the terrestrial landscape, most of the continental shelf has now been mapped in similar fashion. Instead of contouring topography (the elevation above sea level), **bathymetry** (the elevation below sea level) is charted. These charts are readily available and are used by weekend yachters, folks fishing and professional boat pilots alike. On the bathymetry charts, the contour lines are equally spaced at intervals labelled on the chart itself under the heading **contour interval**. The contour interval is frequently found near the linear scale on the chart. Any unit can be used but frequently they will be in multiples of tens of meters, tens of fathoms or tens of feet.

The smaller the contour interval the more detail will show up on the chart. However, at some point, detail may exceed data or the chart will begin to look too cluttered to read easily. Therefore, the contour interval is based upon the relief within the chart area. High relief requires a large contour interval, while a much smaller interval can be used in low relief seascapes. A contour interval of 100 meters would show no information for a bay with a maximum depth of 98 meters and would be quite useless. Even though there may be great jumps in relief from 1 to 98 meters, the relief would not show upon such a chart. Figure 5-5. To assist the chart users, every fifth contour line is presented in bold face. These are known as **index contours**; superimposed on these index contours is the number of units that line represents. Some charts only emphasize a 1.5 or 10 fathom interval line and frequently this is a dashed line.

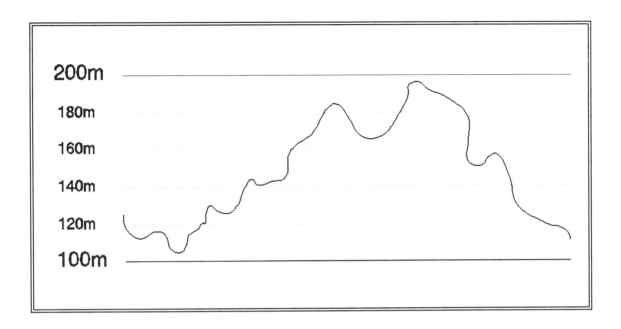

Figure 5-5. Schematic showing array of relief missed by a contour interval of 100 meters that could be readily seen with a 20 meter interval.

Making a Bathymetric Chart

To make a bathymetric chart, one must define the area of interest and determine the maximum relief found with the boundaries. Relief can be determined in several ways depending on the depth of the area. For areas where depth exceeds several hundred meters, usually a form of sonar is used. For our purposes, we will assume a very low tech operation and revert back to the days of Samuel L. Clemens (Mark Twain) when a person at the bow of river boats traveling down the Mississippi River would throw out a knotted line with a weight on the end into the water. When the weight hit the river bed, the person would read off the number of units of depth from the rope on the weight and shout out the depths to the pilot. The name Mark Twain is even from one of these depth readings. We will assume that a similar process can be used along some coasts of interest to us when we create our bathymetric charts in the exercises for this lab.

Once the relief has been determined (in the case of exercises 2 and 3 this is 0-60 meters), the number of contour lines desired to provide maximum information with minimal clutter is calculated. Until one becomes proficient at contouring, it is recommended that the total number of contour lines in any chart you create should not exceed five or six. Dividing the relief by the number of lines desired will give a suggested contour interval. Thus, in exercises 2 and 3, 60 meter/6 lines= 10 meter intervals. This number should be rounded off to the nearest multiple of five or ten for ease in reading.

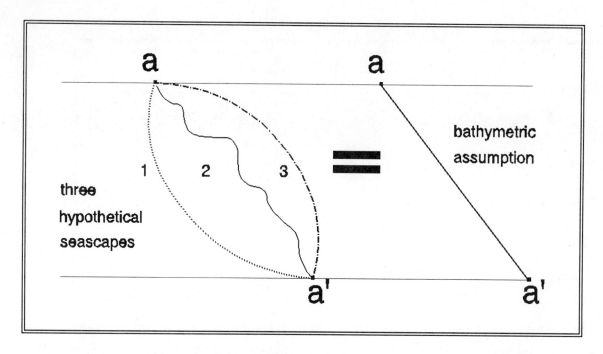

Figure 5-6. Regardless of how uneven the land or seascape is, a uniform slope is assumed between points of known elevation unless data on the map shows otherwise.

The depths for the charts you will make will be provided for you. Not all depths will appear in multiples of the contour interval you have selected. Therefore, the points that appear on the contour lines may have to be derived by interpolation. To do so, assume a slope of uniform gradient between any two adjacent points of known data. Figure 5-6. This means that a point 5/6's the distance between a data point of 95 units depth and a data point of 101 units depth would be equal to 100 units in depth. If 100 units was a multiple of the contour interval selected for this chart, then this newly determined point of 100 units depth would appear on the contour line for that depth. Uniform slopes seldom exist in nature; however, this assumption will help us get started. In reality, if a data point becomes confused and suspect due to interpolation, and if it is also an important position, then it is best to return to the field for an actual measurement.

After data points of multiples of the contour interval have been created between field measured data points, it may become necessary to add additional information points before contour lines are actually drawn. It may not even be possible to recognize places where more information is needed until the contour line positions have been drawn. To add more information you may need to use some of the interpolated points (like the one for 100 units depth determined in the above paragraph), and further interpolate from these points to other field measured points. You may even have to use two interpolation points. Because theoretically every point on any contour line is equal in elevation, any point on a line you have already constructed may be used as an end point for further interpolated.

Once it seems that there are enough points available, the contour lines may be drawn in a similar fashion as a child does when they do "connect the dot activity books". Just as in pictures made in these children's activity books, the pictures created may at first appear blocky, the final version, to really look correct, may have to have some of the lines rounded or smoothed out (Figure 5-7).

Figure 5-7. "Connect the dots" picture - direct from the dots and corrected by rounding out areas for more pleasing presentation.

In the children's books, this is usually no problem as most of us will recognize the image being drawn and "correcting" it is no problem. What about smoothing out or "correcting" contour lines? Since this is a new chart and we do not know what the real area looks like, how can it be smoothed out without losing accuracy? These are good questions and the answers may frustrate some of you. The best way for cartographers to smooth the lines is to have a lot of experience in the field seeing how landforms and seascapes really look and also in drawing charts such as these, so that they recognize patterns which are frequently repeated. Does this mean that the cartographers do not make mistakes in rounding off features? Of course not. Mistakes are corrected on future charts as they are discovered or new data is provided by more field work. The sea floor

Figure 5-8. These charts show the major bathymetric features of the ocean basins. **(From P. Pinet.** *Oceanography: An Introduction to the Planet Oceanus.* **West Publishing Company, 1992.)**

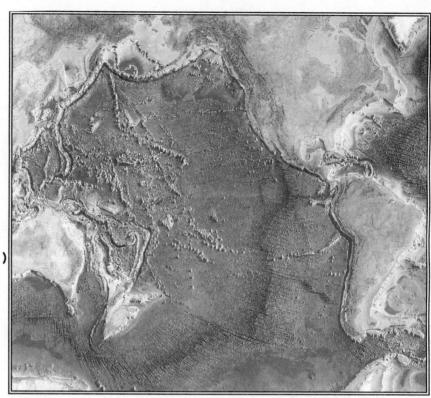

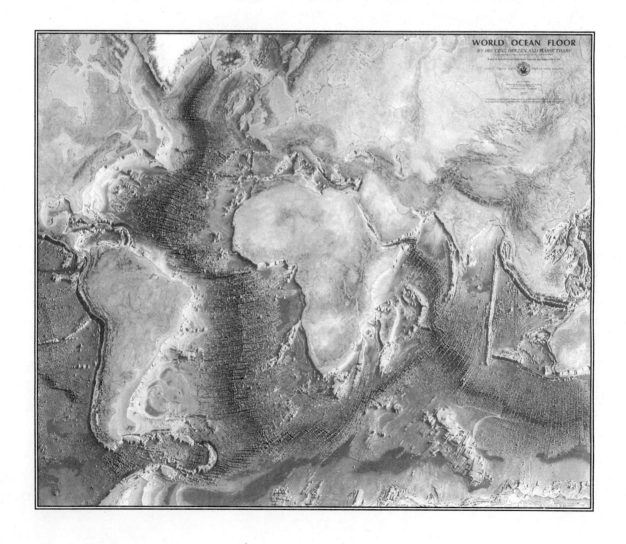

in many instances is not smooth at all but can actually be quite rugged regardless of how large the contour intervals may be. Figures in 5-8 shows the bathymetry of four selected areas of the sea floor.

Topographic and Bathymetric Profiles

As discussed in the above section, topographic maps and bathymetric charts provide for three-dimensional information on a flat surface. This information can be very useful, but it is often more beneficial to see a side view of the features on the chart as well. The information on the chart can easily be transferred into this format. The resulting image is called a **topographic** or **bathymetric profile** or **cross-sectional view**. The profiles can be made quite easily along any straight line on the chart's surface by following the format below.

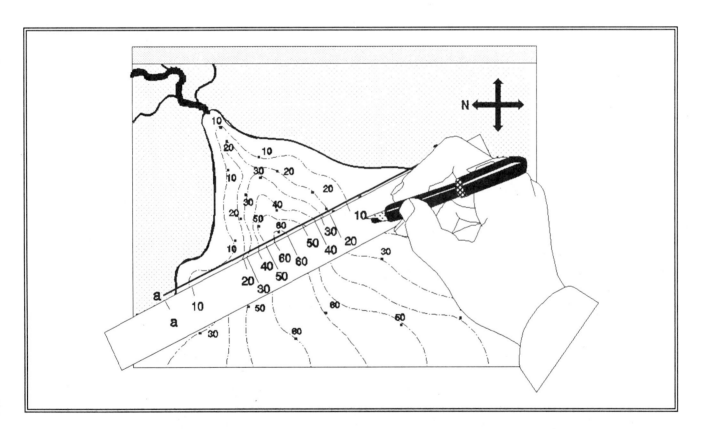

Figure 5-9a. Constructing a bathymetric profile, step 1.

1. Lay a straight strip of paper along where the profile is to be constructed. Mark the end points of the line on the paper strips.

2. Mark on the paper strip the exact place where each contour line crosses the profile line. Even if it is the same contour line crossing multiple times, each time should be marked.

3. Label each mark with the contour level it represents. If the
   lines are closely spaced, it may be sufficient to label only the
   index lines (those that appear as bold or dashed lines on the
   chart).

4. Prepare a vertical scale on the profile or graph paper. This
   scale need not be the same as the horizontal scale of the chart.
   As a matter of fact, often the horizontal scales on the chart may
   be so large that relief of several hundred units may show up only
   as small squiggles on your profile unless vertically exaggerated.
   Vertical exaggeration is calculated by dividing the distance of
   one horizontal unit by the distance of one vertical unit.

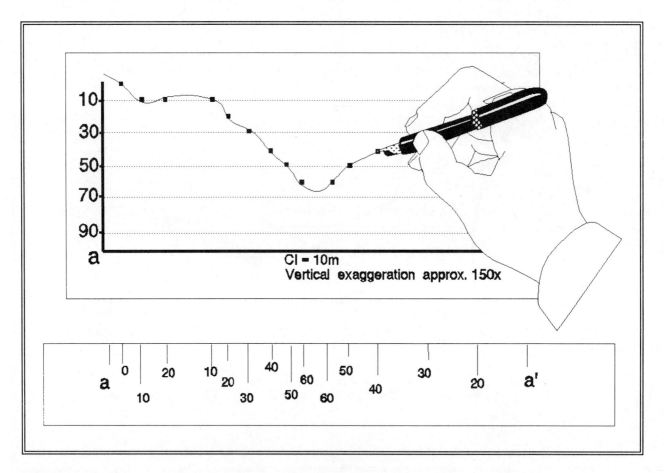

Figure 5-9b. Constructing a bathymetric profile, step 2.

5. Place the paper with the labeled marks at the bottom of the
   profile paper and project each contour onto the horizontal
   reference line that corresponds to that elevation.

6. Connect all the points with a continuous smooth line.

Exercises

1. Contour lines.

In groups of 2-5, create "seascapes" out of oil-based modelling clay in the clear plastic trays provided for you. The seascapes should contain areas of hills and valleys. It is best for our purposes if there are no isolated basins or volcano calderas, i.e. there should be ways for all high places to drain completely. Once your sea scape is created, have it approved by your instructor. This is to insure that your seascape will yield the maximum information on contouring. Fill the tray so that the entire seascape is under water. Mark with waxed crayon along a vertical line on the side of your tray 5-7 equally spaced intervals. The spacing of these marks will be equal to the contour interval for the chart you will create (one-half inch works well).

Drain, by syphon hose, the water level one interval spacing at a time. Between each draining, take a sharp object (a pencil or bent paper clip works well), and trace along the water line on the exposed clay. Continue this procedure until the tray is drained and your seascape is entirely exposed. Discard water down the drain.

Place the tray on the floor and view it from directly overhead, a bird's eye view. On the piece of paper provided, draw the landscape by reproducing the water line tracings only. Notice that your lines follow all the rules of contouring addressed earlier.

Bathymetric Chart of Created Seascape

2. **Contouring Bathymetric Charts**

   a. On the chart labelled Exercise #2-a. No Hope Bay, California, (a mythical location south of San Diego), the depths that are necessary for creating a bathymetric chart for the area have been predetermined. Create the bathymetric chart.

   b. On the chart labelled Exercise #2-b. Sad Sailor's Sound, Florida, (another mythical place), data points are not in even spacing, nor will they all appear on your contour lines. You will now need to interpolate points to create a bathymetric chart of the area. Do so as described in the introduction; then create the bathymetric chart.

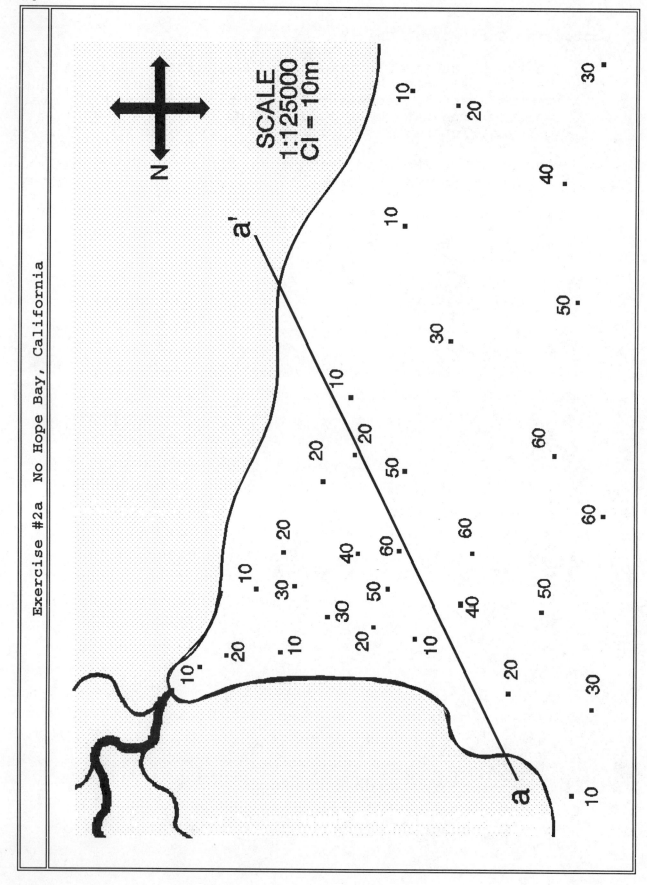

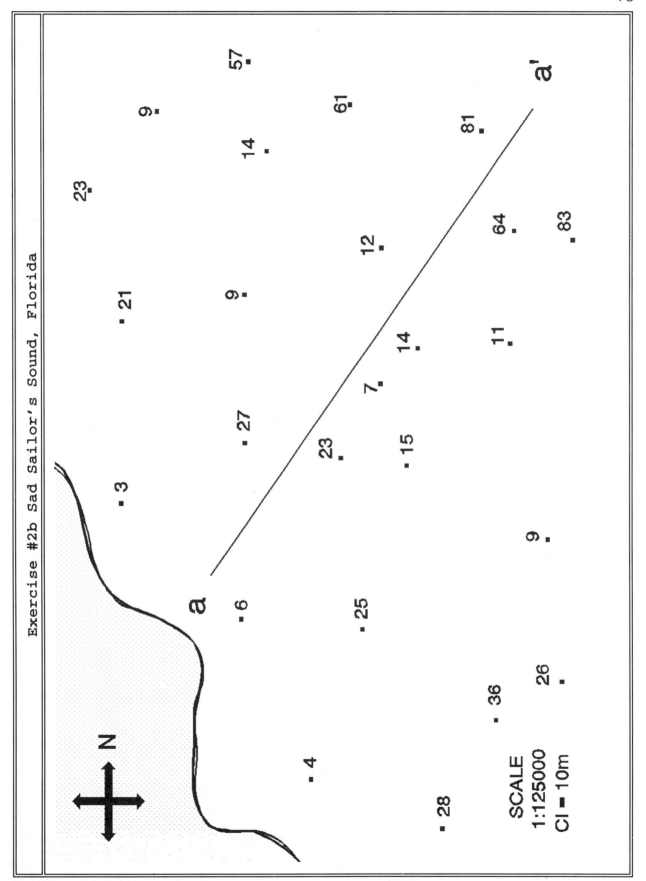

3. **Profiling Bathymetric Seascapes**
   The lines marked a-a' denote profile lines of interest on each of the charts. Draw profiles for each using the method described in the discussion above.

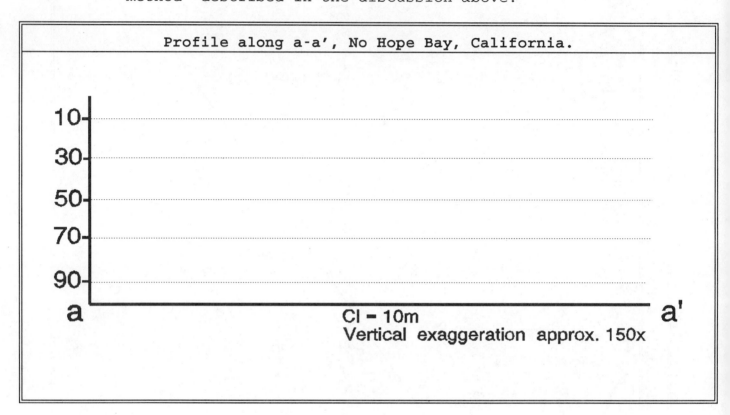

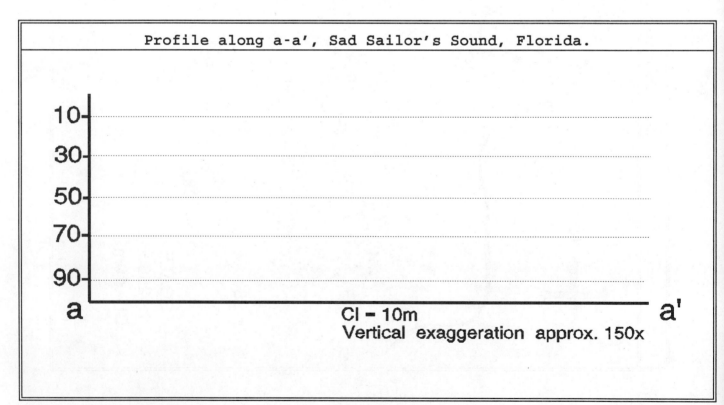

4. **Questions on Bathymetric Charts and Profiles**
   Answer the following questions for each of the charts based on your created bathymetric charts and profiles.

   a. What feature(s) do you see in your charts? Are there any high or low points?

   b. How do you believe they may have been formed? Are they linear or do they show a closed geometric form? Are the features oriented in any particular manner?

   c. Are there any physical changes in sediments size or in sediment composition that may help you recognize how the features formed? Explain.

Answers to question #4 - Chart 2A.
a.
b.
c.

Answers to question #4 - Chart 2B.
a.
b.
c.

d. What problems did you encounter in contouring the charts?

Answer to question #4d

e. In what specific areas did you have a problem interpreting the bathymetry?

Answers to question #4e

f. What information would you require to improve your diagrams or your interpretations?

Answer to question #4f

Glossary of New Terms

Topographical Map: Maps showing relief on land by the use of contour lines.

Contour lines: Lines connecting points of equal elevation on topographical maps and bathymetric charts.

Bathymetric charts: Maps showing submarine relief by the use of contour lines.

Relief: The difference between the highest and lowest elevations in an area.

Contour interval: The vertical spacing between contour lines.

Index Contours: Every fifth line on a contour map at common multiples, darkened or dashed for quick reference.

Profile or cross-section: A side view of the feature in a topographic map or bathymetric chart.

Further Readings

Carr, A.P. The ever-changing sea level. *Sea Frontiers*. 20(2): 77-83, 1974.

Emery, K.O. The continental shelves. *Scientific American*. 221(3): 106-25, 1969.

Heezen, B.C. The origin of submarine canyons. *Scientific American*. 195(2): 36-41, 1956.

Menard, H.W. The deep-ocean floor. *Scientific American*. 221(3): 126-45, 1969.

Schafer, C., and Carter, L. Ocean-bottom mapping in the 1980's. *Sea Frontiers*. 32(2): 122-30, 1986.

> *And Thou, vast Ocean! on whose awful face Time's iron feet can print no ruin trace.*
>
> Robert Montgomery, *The Omnipresence of the Diety*, 1828

Laboratory SIX

Coastal Formation

Coastlines and Coastal Formation

Coastlines, the interface between land and the sea, have been formed by a variety of processes. By their very nature, coasts are in constant change because they act as an energy buffer as well as a transition between the terrestrial and marine worlds. When most people hear the word coast or shoreline, the images they recall are of coconut palm-lined sandy beaches. Although this picture is certainly appealing, this notion is far from typical of most of the world's continental or island margins. Beaches account for only about 10% of the shoreline morphology.

Today, most oceanographers and geomorphologists (geologists who study landforms) classify coastlines on the basis of the processes that formed them. **Primary coasts** are those that have been shaped principally by terrestrial or land-based actions such as running water in rivers, ice in the form of glaciers or the creation of "new" rocks (volcanism). **Secondary coasts** are those which have been principally formed by ocean-based processes such as waves, tides, or coral reef development. Each of these two main groups can be subdivided into **positive** forms caused by deposition of material, and **negative** forms caused by erosion. Another classification system uses the concepts of submerging or emerging coastlines as the major division of classification. Each system has its strengths; however, we will use the former for our purposes.

Regardless of the classification, nearly every coast has, in recent years (the last 50 thousand or so), been greatly affected by the relative changes in sea level that have occurred as continental glaciers melted and retreated, then refroze and advanced during

Pleistocene Time. Since we live in a closed system with respect to water, the water needed to form the miles-thick ice covering the northern and mid-continent areas of North America and the rest of the higher latitudes of the world during the "Ice Age" must have come from somewhere. The logical place is the oceans.

It takes only elementary mathematics to estimate the amount of ice that covered the continents during times of maximum glaciation (if one assumes even a modest uniform ice thickness). With a little fancier calculation based on the weight needed to deform or depress the strata that we know was depressed by the ice in some localities, more accurate thickness estimations can be made. It is also a simple calculation to remove the amount of water needed to form these amounts of ice from the oceans and see where, on a topographical map, the "new" low standing water boundary would be. Obviously, the boundary would be somewhere that is currently ocean-covered. We will do a similar experiment in a later lab when the topic of sea level, its fluctuations and the causes of such drastic fluctuations will be addressed in greater detail.

Does this mean that if we go out in a submarine we might see ancient shorelines and beaches on the continental shelves at some depth? Perhaps, but certainly over the past few tens of thousands of years many of these drowned features would have been covered by ocean sediment to a point where we may not even recognize them as the features they once were. Some are still quite obvious, however.

What about the features formed when the sea levels were higher than they now are? Can we see these ancient coastal features on land where deposition isn't so rapid and erosion often takes longer than even tens of thousands of years? The answer is definitely yes! These features are often quite obvious today. A classic example will be seen among your exercises in lab on the maps of southern California. One other reason some of these areas have been left high and dry is that, at some localities, the former coastlines have risen out of the water. We will discuss this possibility later.

Exercises

1. **Analyzing Coastal Morphology**
    We will be using topographic maps produced by the United States Geological Survey. These were explained in detail in the last chapter on topography and bathymetry. Common contour intervals on maps of areas only a few miles in extent are often 10, 20, or 50 feet. Maps of very large areas or mountainous areas may have contour intervals as great as 200 or even 500 feet. As seen in the last lab, the greater the contour interval the more detail will be missed on the map. Remember, the **contour interval** can be seen near the scale usually centered at the bottom of the map.

    a. Working in pairs or teams of pairs, view the maps of the coastal areas provided. Utilizing all of your past experiences, try to determine where each map fits the classifications of **primary or secondary** and **negative or positive**.

    b. Write a paragraph explaining why you believe the area shown in each map fits the description you have assigned to it. If you work in groups, only one lab write-up needs to be turned in, but each group member should sign it and is responsible for the report's content.

    Once again you are not to run to a reference book and find "the answer" to this lab exercise. It was designed to encourage use of your reasoning and observation skills. You will be graded on how well you support your responses, not if your answer agrees with the experts.

EXAMPLE:
Map X) Sad Sailor Sound, Florida.

Secondary, Positive.

This is an interpretation of the bathymetric chart created in exercise 2b of Lab 5 on Bathymetry. Refer to it as needed. The build-up seen running parallel to the coast several hundred meters offshore has been produced as a biological reef and has been confirmed by the reefal limestone cores taken recently. Coral reefs are a marine phenomena; therefore, the secondary classification is appropriate. Since the feature is depositional, the positive classification seems best.

EXERCISES
MAP A   Terrebonne Bay, Louisiana
MAP B   Morehead City, North Carolina
MAP C   New Bedford, Massachusetts (Map and Landsat) Do the different methods of showing this area reveal added information useful for interpretation? If so, how?

MAP D	Point Sur, California

MAP E	Bath, Maine (2 maps/different scales)
Do the different scales in these two maps reveal different trends or alter your interpretation? If so, how?	

MAP F	Redondo Beach, California

MAP G	Elizabeth, North Carolina

MAP H	Eureka, California

Glossary of New Terms

Negative Feature: Landform created by erosion.

Positive Feature: Landform created by deposition.

Primary Coast Development: Shorelines caused principally by land-based processes such as rivers, glaciers, etc.

Secondary Caost Development: Shorelines caused principally by ocean-based processes such as waves, current, tides, etc.

Further Readings

Bascom, W. *Waves and Beaches: The Dynamics of the Ocean Surface*, rev. ed. Garden City, N.Y.: Doubleday Anchor Goods, 1980.

Bird, E.C.F. *Coastlines*. Cambridge, Mass.: MIT Press, 1969.

Carefoot, T. *Pacific Seashores*. Seattle: University of Washington Press, 1977.

Darwin, Charles. *The Structure and Distribution of Coral Reefs*. Berkeley, California: Univ. of California Press (a 1962 reprint of the original), 1842.

Feazel, C. The rise and fall of Neptune's kingdom. *Sea Frontiers* 33(2): 4-11, 1987.

Fulton, K. Coastal retreat. *Sea Frontiers*. 27(2): 82-88, 1981.

Gosner, K.L. *A Field Guide to the Atlantic Seashore*. Boston, Massachusetts: Houghton Mifflin Co, 1979.

Hopley, D. *The Geomorphology of the Great Barrier Reef*. New York: Wiley Interscience, 1982.

Kaufman, W., and O. Pilkey. *The Beaches Are Moving*. Garden City, New York: Anchor Press/Doubleday, 1979.

Leatherman, S.P. *Barrier Island Handbook*. Amherst, Massachusetts: Univ. of Massachusetts Press, 1979.

Shepard, F.P. and H.R. Wanless. *Our Changing Shorelines.* New York: McGraw-Hill, 1971.

*Why was the sea made salt?
Because, I think
If fresh, the fishes every drop
would drink.*
         Timotheus Polus (d. 1632)

Laboratory SEVEN

Marine Chemistry

Dissolved Solids and Salinity Determination

Perhaps the most obvious difference in water from the oceans and water from most lakes or rivers is its salt content. Rivers contain, on average, 0.01% (or one part per thousand) of dissolved salts while, on average, seawater has 3.5% (or 35 parts per thousand) of various salts. Any water that has come in contact with the earth's surface or the atmosphere contains dissolved salts; even rain and snow have some small amounts. The relative amount of solids dissolved in water has several effects. Even though not all the dissolved materials in seawater are salts, as most of us envision salts, the total amount of dissolved solids in water is referred to as **salinity**. In spite of all the compounds that can be found in the ocean, a mere six elements make up over 99% of the dissolved solids in the water of the seas. These six elements (ions) and their relative concentrations are listed in Table 7-1. Increased salinity has several effects on water properties; a few include:

o   Lowing the freezing point and raising the boiling point

o   Increasing density

o   Decreasing vapor pressure

o   Increasing osmatic pressure (important in marine and freshwater organisms).

Ion	Parts per 1000
Chlorine, Cl^-	55.05
Sodium, Na^+	30.61
Sulfate, SO_4^-	7.68
Magnesium, Mg^{++}	3.69
Calcium, Ca^{++}	1.16
Potassium, K^+	<u>1.10</u>
Total of 6 =	99.29

Table 7-1. Major constituents of dissolved solids found in the world's oceans and percents by weight.

In addition to the dissolved solids there are also gasses contained in seawater. Oxygen, nitrogen and bicarbonate ($HCO3^-$) are by far the most common with a host of others making up a total of less than 1% of the dissolved gas content. Surprisingly, the ratio of the material dissolved in seawater, by percent weight, is relatively constant regardless of the salinity. This observation is important in a number of respects including the fact that it provides one way to measure the salinity of water. The consistency of the composition makes it necessary to measure only one of the six major elements. Because chlorine is most abundant and easily measured, the chlorine ion (Cl^-) is most frequently used. Although not mathematically absolute, oceanographers have chosen the constant 1.80655 to multiply the percent weight of Cl^- in a seawater solution to determine the salinity. An international organization, now located in England, provides samples of standard seawater used universally to calibrate equipment used to measure salinity.

Most seawater is somewhat basic or alkaline with ranges between 7.5 - 8.5 on the Ph scale. When the carbon dioxide (CO_2) in the bicarbonate dissolves, it forms a weak acid (H_2CO_3) which helps buffer, or neutralize, the alkalinity. It may be difficult to understand how the carbon dioxide comes and goes in the water because it is a complex cycle. Carbon dioxide is produced by decaying plants and animals on the sea floor. Some CO_2 entering the sea also comes from the atmosphere via the respiration of plants and animals. Carbon dioxide can be readily altered and locked up as calcium carbonate ($CaCO_3$), the main component in shells, corals and bone tissue. Most of the calcium carbonate in the ancient seas is currently contained in limestone deposits worldwide. Throughout the North American Midwest, limestone deposits can be found that are hundreds of meters thick. The limestone is from the shells, corals and bone tissue being deposited on the sea floor and incorporated into the sediment cover.

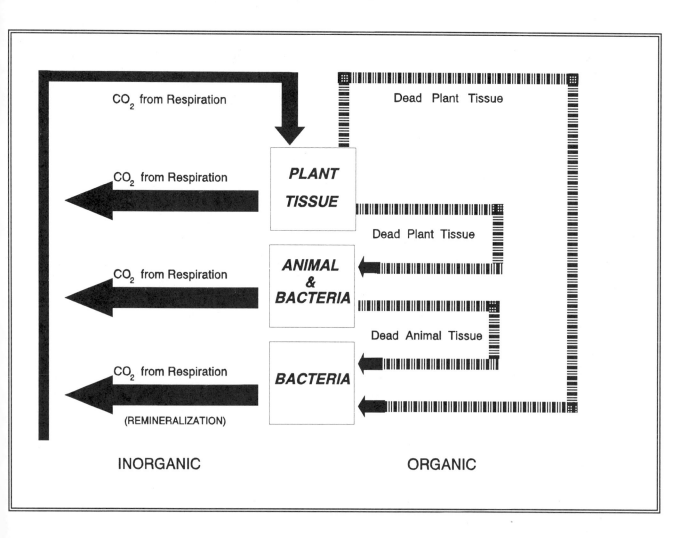

Figure 7-1. The Biogeochemical Carbon Cycle in Ocean Water.

The process of CO_2 being dissolved in seawater and being used by organisms to create hard body parts has been going on as long as organisms have had hard parts, at least 1 billion years. The $\underline{C}$ in CO_2 is carbon, the $\underline{O}$ is oxygen. In trying to follow the cycle of CO_2, carbon and not oxygen is usually monitored, Figure 7-1. Carbon is much easier to trace than oxygen which is seemingly everywhere in the seas. One may wonder how long the carbon stays in the loop or cycle; a few days? weeks? years? centuries? The time any element or compound stays in the cycle is called its **residence time**. A better way to think of it may be the time it takes to entirely replace, by natural means, any given substance in the ocean. Some elements stay locked up for so long that it may as well be an eternity. Sodium and chloride, the two elements in table salt, both have residence times over 65 million years! Carbon, by the way, has a residence time of around 8,000,000 years.

Exercises

Today you will determine the salinity of seawater from the same source by a variety of means, then compare results of the methods used.

1. **Calculating Salinity I**
    One of the areas of difficulty many people who are not from coastal areas have is imagining the amount of material dissolved in seawater. To assist your understanding, weigh an empty container (provided). Fill the container with water from the ocean and reweigh. Over the next several weeks of the course, the water will be left to evaporate. This may have been done already in class. After the water has evaporated completely, the container will be weighed again, and the percent of dissolved solids by weight can be determined.

$$\frac{(weight\ of\ residue + container\ weight) - container\ weight}{(weight\ of\ water + container\ weight) - container\ weight} \times 100 = salinity$$

Answer to Question #1
Weight of empty container:
Weight of container + water:
Weight of container + residue:
Salinity (in %):

Do not throw the residue away. You will use it later.

2. **Calculating Salinity II**
    Some additional water from the first exercise was saved for your use at this time. One of the observations mentioned in the earlier discussion is that as salinity increases so does the density of the seawater. The relatively simple instrument used to measure density of fluids, especially those with densities near 1, the specific gravity of water, is the **hydrometer.** Use a hydrometer to measure the salinity of the seawater in lab. The value of the density can be read directly from the instrument as long as the density is within the range that the hydrometer will readily measure. If the reading is at the maximum level of the instrument, then a hydrometer with a broader or higher range should be used to check your results. This technique is not very accurate, but it is fast, relatively simple and, since a hydrometer costs less than $10.00, inexpensive.

3. **Primary Metal Ions in Seawater**
To determine the primary metal ions in seawater, take some of the residue from the evaporated seawater from exercise 1, and powder it. A mortar and pestle work well for this process. To determine the primary metallic constituents of the residue, take a clean metal spatula with the residue powder on it and place it over a flame from a bunsen burner or other source. The flame will quickly flash a bright color. The color in this case should be a bright yellow, indicative of the presence of sodium. Other colors are present but are masked by the high ratio of sodium present (Table 7-1). Although older books indicate that chlorine will flash blue, chlorine is not a metal and will not flash brightly. There are several methods using advanced equipment that can do a more precise determination of elements present. The same principle of matching the colors of the ignited elements to charts or standards of known element emissions is utilized in many tests. Some tests can even determine the relative amounts of the element present. The observation that different metals burn with different colors has been used quite successfully by **pyrotechnists** for centuries. Pyrotechnists create the fireworks for aerial displays that are so popular on Independence Day.

Other methods of determining salinity using the **index of refraction** (the property of light traveling and bending through transparent substances) and the **conductivity** of the seawater can be demonstrated to you in class if your institution has equipment to do so. Conductivity is the measurement of how readily an electrical current can pass through a substance. In water, conductivity is directly related to the concentration of dissolved solids. This method of salinity determination is used most often by professional oceanographers in the field.

4. **Residence Time**
Calculate the residence time of water in the ocean using the following equation:

$$Residence\ Time\ in\ years = \frac{amount\ present\ in\ the\ ocean}{rate\ of\ removal\ per\ year}$$

To determine values for the variables for this exercise:

o Return to the hypsometric curve in lab chapter 2 and determine the average depth of the ocean. Figure 2-1.

o According to the National Oceanographic and Atmospheric Administration (NOAA) there is a net removal of water due to evaporation of about 10 cm per year. Most of the evaporated water finds its way back to the sea eventually so the level changes only slightly year to year. This will be discussed in greater detail in the lab on sea level as a datum.

o   Be sure to keep track of decimal point location!

Answers to question #4
Depth of the ocean:
Residence time in years:

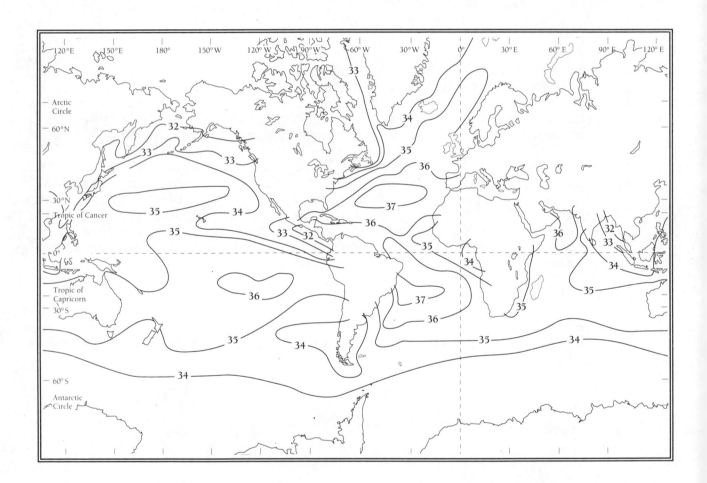

Figure 7-2. Sea surface salinities in parts per thousand for the month of August. (From H.U. Sverdrup, M.W. Johnson and R. H. Fleming, *The Oceans*, 1942).

5. **Salinity Gradients**
    The pattern of salinity ranges at the sea surface shows a very high correlation with latitudinal bands. Figure 7-2. Maximum values for salinity occur in two geographical bands: 1) between 15-20o south latitude, and 2) between 20-30o north latitude. Lesser values are found nearer the equator with even lower salinities at the poles. What may be the cause of such zonation?

Figure 7-3 is a map of sea surface temperatures in August. How might this map be related to the one in figure 7-2 (average salinities at the ocean surface in August)? Although the detailed answer to this question is quite complex look for generalities. How does evaporation, ice production, river run-off and ice melting affect salinity locally, regionally, globally?

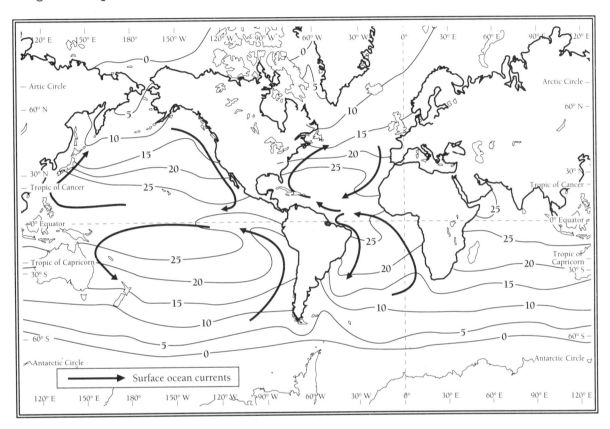

Figure 7-3. Sea surface temperature averages in degrees centigrade for month of August. (From H. U. Sverdrup, M. W. Johnson, and R.H. Fleming, *The Oceans*, 1942.)

Answer to question #5

Glossary of New Terms

Conductivity: The measurement of how readily an electrical current can pass through a substance.

Hydrometer: Instrument that measures densities of liquids near those of water. Because water increases density with increased salinity, this instrument can be calibrated to read salinity directly.

Index of Refraction: The property of light traveling and bending through transparent substances.

Pyrotechnists: Technicians who create the fireworks for aerial displays.

Residence Time: The average time any element stays in the sea during its cycle.

Salinity: The total amount of dissolved solids in water.

Further Readings

Broecker, W.S. *Chemical Oceanography*. New York: Harcourt Brace Jovanovich, 1974.

Harvey, H.W. *The Chemistry and Fertility of Sea Water*. 2nd ed. Cambridge: Cambridge University, 1957.

Hedgepeth, J.W. (ed). *Treatise on Marine Ecology and Paleoecology*, v. 1, Ecology Geological Society of America, Memoir 67, 1957.

Pinet, P. *Oceanography: An Introduction to the Planet Oceanus*, West Publishing Company, 1992

Shepard, F.P. *Geological Oceanography: Evolution of Coasts, Continental Margins and the Deep Sea Floor*. New York: Crane, Russak, 1977.

Sverdrup, H.W., M.W. Johnson, and R.H. Fleming. *The Oceans*. Englewood Cliffs, New Jersey: Prentice-Hall, 1942.

Laboratory EIGHT

Waves

Measuring Wave Parameters

Nearly all of us have been in awe of the power of water waves at some time or another. Whether it was while watching old movies on "tidal waves", TV coverage of hurricanes, or even in person at the sea shore or perhaps the shore of one of the Great Lakes during a storm, most of us have seen waves in their heights of glory. News coverage tends to accentuate the destructive nature of waves, but waves serve an extremely important function in the balance of beach budgets as we will see in later labs. As we will also see, man's efforts to control waves and other natural forces are extremely expensive and will ultimately end in failure every time.

Waves can be created in a number of ways. Most of us have thrown rocks into lakes and watched the concentric ripples flow outward from the point of entry. Although there are occasional meteorites striking the ocean, this form of wave production is relatively rare. Most ocean waves are created by the wind moving over the water and transferring the energy of the wind's motion to the water by friction. Although waves start out small as the wind begins to blow, hugE waves can result. Unidirectional, sustained winds of 70 miles per hour, over large areas, have been known to produce waves nearly one hundred feet high. Waves continue to grow as long as the winds stay strong and continue blowing in the same direction. Wave growth in such cases is limited primarily by gravity and the length of open sea available (**fetch**) before the waves crash onto the land.

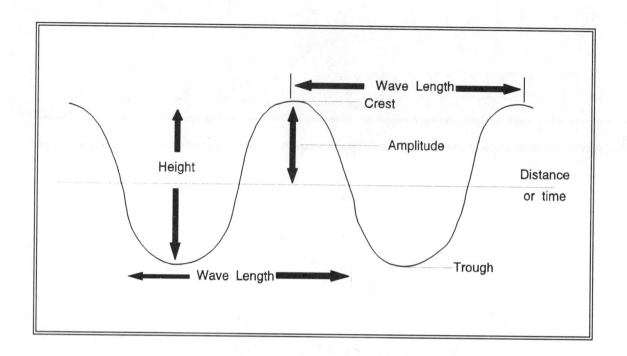

Figure 8-1. Wave parameters.

Waves are measured in a number of ways. The most frequently used parameters are the **height** and **length** of the wave (Figure 8-1) and the **wave period**; the length of time in seconds that it takes for one complete wave length to pass a fixed point. One other measurement often used is **frequency**; the number of waves or wave fractions passing a single point over a given duration of time. If one second is used as the time interval, the measurement for frequency is given in **hertz**. You may have heard this term over the radio when the station identifies itself. The term hertz is used as a measurement of many wave phenomena.

In the deep sea these apparently simple parameters can actually be somewhat difficult to measure. Wave height and period can be readily measured, but wave length is much more difficult without the use of photographs. Periods for waves are measured as the crests of the waves pass a fixed object like a pier, pole or, in the open ocean, a fixed reference like a large vertical floating object. To determine wave periods, usually one begins timing the waves as a crest passes the given fixed point and ends as the eleventh crest passes the same point. The time duration of the ten waves measured is then divided by ten for an average period per wave. When visual observation is not possible or practical, mechanical devises that use pressure-sensitive pads aligned in a vertical line are substituted.

Ideally, the waves in the sea would have a uniform sinuous shape like the ones shown in Figure 8-1. However most sea waves are quite complex with multiple or peaked crests and have smaller waves superimposed on the larger ones. The waves found in the open ocean are combinations of waves that have frequently been created from multiple source areas. Waves of uniform size and speed are somewhat rare.

Many devices have been created to reproduce waves in the laboratory and today you will be introduced to one of them. The apparatus for the focus of our attention is a wave table or tank (Figure 8-2). This device allows us to create waves and monitor their effects. We will do this in several subsequent lab exercises. The table also allows us to see how interference waves are formed and how waves react to various obstacles both natural and manmade (cusps, spits and breakwaters, etc). In this lab we will concentrate only on the waves themselves and how the water in the waves moves.

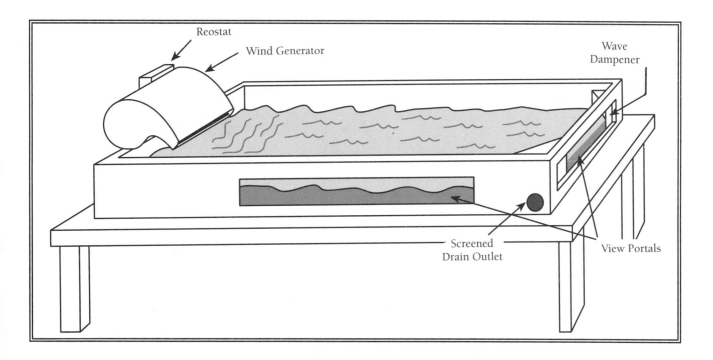

Figure 8-2. The wave tank and its parts.

Ocean waves are the transfer of energy through water. Waves are not the movement of the water itself. If you have ever been fishing or boating you may have seen objects floating on the lake bob up and down or move in a rotary motion as waves pass by. If you watched closely the object stayed more or less in the same place. Figure 8-3.

Waves can be divided into three categories based on the relationship of the ratio of the wave's length to the depth of the water. **Deep water waves** are those whose length is greater than one half of the water depth. **Shallow water waves** are those in water shallower than 1/20 the wave's length. **Intermediate** or **transitional waves** are those that fall between these depth/length ratios.

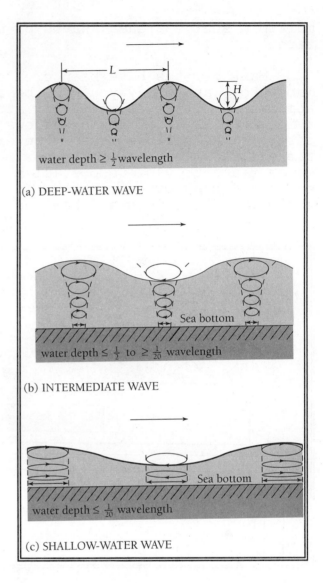

Figure 8-3. The motion of an object as waves pass in deep water, intermediate water depth, and shallow water. (From P. Pinet, *Oceanography: An Introduction to the Planet Oceanus,* **West Publishing Company, 1992.)**

The energy from deep waves diminishes proportionally to a point of nearly zero at about the depth of one half the wave's length. Thus, by definition, energy from deep water waves never effects the sea floor.   Figure 8-4.

The depth of everyday wave energy and its effect on the sea floor is used to define an area of the coastal environment known as **normal wave base**. This represents the area near-shore where sediments are in nearly constant motion in response to energy from the surface. A deeper reference plane can be delineated where only the most severe storm waves, with very long lengths, affect the bottom. This area is known as the **storm wave base**. Figure 8-5 shows the coastal zonation by these parameters. The depths of these zones, will vary locally depending on the regions climate and **fetch** (length of open sea that can be effected by the wind that creates the waves).

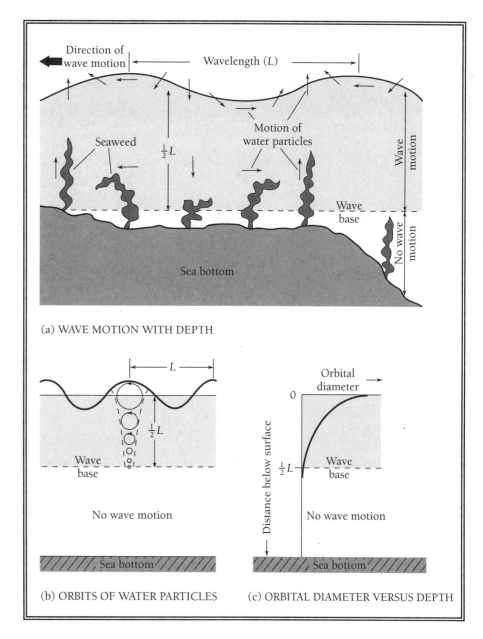

Figure 8-4. Orbital paths of wave particles beneath waves. (From P. Pinet, *Oceanography: An Introduction to the Planet Oceanus*, West Publishing Company, 1992.)

As the deep-water waves near shallower areas, the orbital movement of the objects and water in the waves begins to flatten out and becomes more back and forth in its motion. Figure 8-3. The friction along the sea floor increases and the wave's base begins to drag. As the wave's crest continues nearly unaffected it soon outruns the base and has nothing to support it. The crest then falls, forming a breaker or breaking wave. Because waves usually strike the beach at an angle, the portion that nears the shore first will slow down. The waves will then bend to strike the coast nearly parallel to the beach face. Figure 8-5. If waves in an area are nearly uniform in length, then the waves will continue to break at nearly the same location. Figure 8-6. This area is known as the breaker zone and is where surfers frequently try to catch waves just as they begin to curl. We will see other results of these shallow near-shore waves on beaches in a later exercise.

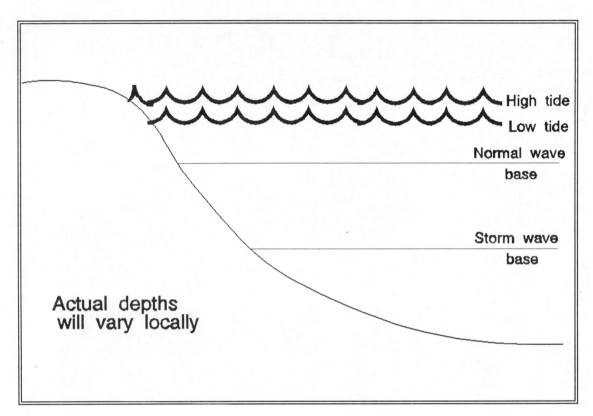

Figure 8-5. Coastal zonation by wave effects on sea floor.

Figure 8-6. Shallow water transformation of breaking waves. (From P. Pinet, *Oceanography: An Introduction to the Planet Oceanus*, **West Publishing Company, 1992.**)

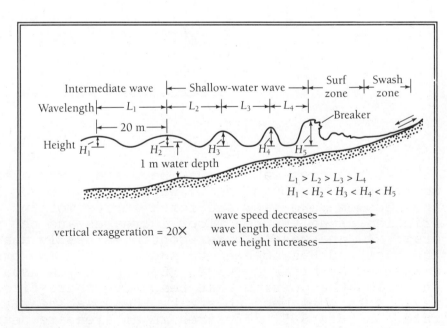

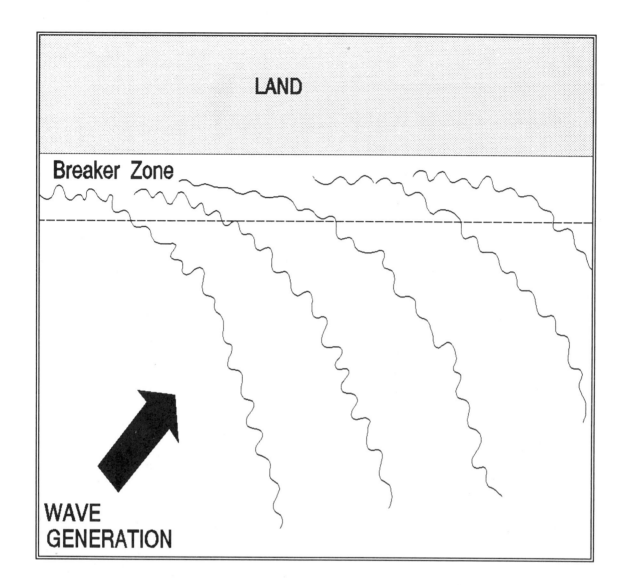

Figure 8-7. The bending of the waves as they approach the shore. As the wave nears a parallel position with the beach face, it will begin to break, delineating the breaker zone.

Exercises

1. **Measuring wave parameters.**
   With the assistance of your instructor, prepare the wave tank so that, if it has a side view window, the at-rest level of water in the tank is about in the vertical center of the view port.

   a. Set the wave maker at the maximum or at a setting that will have the highest wave crest and lowest trough still visible in the window. Leaving the setting constant, measure the height of the wave and attempt to measure wave length. (This may be difficult as discussed earlier, but try.) Record your results. Set a fixed vertical object like a pencil anchored in a plasticine modeling clay base near the view port or draw a vertical crayon line on the view window and measure the waves period and frequency. Write any observations you make about the waves down to refer to during lab discussion time.

Answer to question #1a

   b. Set the wave maker at a lower speed and repeat the measurements. Are there any differences in measuring waves at this setting? Why?

Answer to question #1b

2. **Measuring wave depth.** Measure the water depth in the tank and determine whether the waves measured in Exercise 1 A and B were deep waves, shallow waves or intermediate waves. Show calculations.

    o Deep waves have lengths > 1/2 water depth

    o Shallow waves have lengths < 1/20 water depth

    o Intermediate waves have lengths between shallow and deep waves.

Answer to question #2

3. **Demonstrating energy transfer.** Place a small wooden bead within view of the side window but not next to it. (The friction of the side of the tank itself and the reflection of the waves may alter the path of the bead.) Using the two settings used for exercise 1A and B, monitor the bead's movement at the surface. Notice it is the energy that is being transferred and not the water itself that is moving in the direction of the wave. Do your observations match those in Figure 8-3?

Answer to question #3

4. **Discussion questions.** Assuming that there were uniform winds going around the world uninterrupted where might one expect to find the largest waves? Why? Are these areas frequently hit by large storm waves? How can you find out?

Answer to question #4

Glossary of New Terms

Deep water waves: Waves whose length is greater than one half of the water depth.

Fetch: Surface area of water affected by wind generating waves.

Frequency: The number of waves or wave fractions passing a single point over a given duration of time.

Hertz: A measurement of wave frequency if time duration is one second.

Normal wave base: The area near-shore where sediments are in nearly constant motion in response to energy from the surface.

Intermediate or transitional waves: Those that fall between deep and shallow water waves.

Shallow water waves: Waves in water shallower than 1/20 the wave's length.

Storm wave base: The area where only the most severe storm waves, with very long lengths, effect the bottom.

Wave height: Distance from lowest trough point to highest point on crest of wave.

Wave length: Distance from crest to crest or trough to trough of wave.

Wave period: The length of time in seconds it takes for one wave length to pass a fixed point.

Further Readings

Bascom, W. *Waves and Beaches*, 2nd ed. Garden City, New York: Anchor/Doubleday, 1980.

Bascom, W. *Waves and Beaches: The Dynamics of the Ocean Surfaces*, rev. ed. Garden City, N.Y.: Doubleday Anchor Books, 1980.

Land, T. Freak killer waves. *Sea Frontiers*. 21(3): 13-41, 1975.

Myles, D. *The Great Waves*. London: Robert Hale, 1986.

Pinet, P. *Oceanography: An Introduction to the Planet Oceanus*, West Publishing Company, 1992

Ross, D.A. *Energy from the Waves*. New York: Pergamon-Elmsford, 1979.

Sverdrup, H. U., M. W. Johnson, and R.H. Fleming. *The Oceans*. Englewood Cliffs, N.J.: Prentice-Hall, 1942.

Nothing under heaven is softer or more yielding than water; but when it attacks things hard and resistant there is not one of them that can prevail.
                    A Chinese Scholar, 240 B.C.

Laboratory NINE

Beach Formation

Beach Formation and Longshore Drift

Although only 10% of the world's shoreline is made up of beaches, sand-lined shores are important economic, aesthetic and environmental features of the sea/land interface. Beaches are temporary features that are actually quite fragile and very sensitive to short-term changes such as occasional large storms or seasonal shifts. Beaches are also susceptible to longer term alterations, such as changes in coastal shape due to isostasy, erosion, changes in sediment supply or - perhaps of most importance to our study - man's intervention into the system.

Today we will begin a laboratory experiment that will actually take several hours to complete. Fear not, however; you are not expected to remain the entire time the experiment is running. We return to the wave table for today's demonstrations and exercises. It will allow us to see how waves react to various obstacles both natural and manmade (cusps, spits and break waters, etc.) and to see the natural processes involved in moving sediment along coastal margins.

Demonstrations/Exercises

Demonstration/Exercise #1 - On Shore/Off Shore Seasonal Migration of Sand

The purpose of the first demonstration is to show how beaches react to the seasonality of wave force. Remember that waves are normally the result of winds that blow over the sea surface. During certain times, the atmosphere is more turbulent than others. This turbulence is somewhat seasonal. Most of us have heard of the term "Hurricane Season," and as Gordon Lightfoot sang in the *Wreck of the Edmond Fitzgerald*, "the storms of November" have always been notorious even on Lake Superior. Winter waves, on an average, have a tendency to be much stronger than summer waves. This is clearly seen in beach development. Gentle summer waves move the sand shoreward and build accumulations along the coastal margins. Larger winter waves destroy the beaches compiled during the summer and shift the sand offshore into sand bars parallel to the coast. In either case, as noted in the lab on waves, the beach acts as the location where the energy from waves is dissipated.

Of course waves of both summer and winter are far more complex and compounded than the simple, uniform waves made in the demonstration tank; but the results, as far as the sand budget processes, are quite similar.

The tank has already been prepared in some respects for your lab. The sand has been sieved to a very fine sand size (1/16 - 1/8 mm) previously. This was done to attempt to limit sand size as a variable when the system is scaled to a smaller level. Experience and repeated attempts have shown that sand this size will be transported by the waves produced in the wave tank. Notice that the sand looks brown from a distance but upon closer inspection there are black grains and even some clear or white grains. As we will see in a later lab exercise, sand can be composed of an infinite variety of materials.

Although nearly any sand of the correct size will work, some mixtures work better than others. One mixture found to work well is from the banks of the Mississippi River near Oquawaka, Illinois, and is mostly a quartz based sand. Added to the sieved sand are sieved carbonate sands from the Florida Keys. These were added in about a 20:1 ratio with the river sand dominating. The reason for the mix is that in some experiments it becomes easier to see the results if there is a separation of the grains by density and, in this case, color. The sand normally remains in the tank and we do not change the sand's composition for each experiment or demonstration. This river sand/carbonate beach sand mix has proven to be workable for many experiments but sand size and composition are areas that are constantly being tested for new and better mixtures.

Some wave tanks, like the one used by the U. S. Army Corps of Engineers in Hattisburg, Mississippi, to model the Mississippi River, use very finely crushed walnut shells. The corps believes that the shells, being less dense than even carbonate sand, give a truer sediment model for the scaled down system they operate. To date we have not tried the shells in our tank.

Procedures

1. Create a beach face that extends completely from one side of the tank to the other with the sand in the tank. The "island" stretching from side to side should be about six inches high

(above water level). Because every tank will be slightly different and the wave generators will be likewise different, experience will show where to position the shoreline so that the demonstration works best. In ours we have the front of the ridge about six inches inside the back of the side view porthole. Figure 9-2. Your instructor may have more specific instructions tailored to your equipment. This is perfectly normal. Draw a sketch of your initial set-up or take an instamatic photograph.

2. Start the wave generating device and set it so that the waves have a relatively long length and low frequency. This simulates summer conditions. Sand should visibly begin to move. If there is only a small amount of sand in front of the beach face the experiment will still work, but will require a longer duration. To save time you may wish to introduce more sand to the tank about 6-10 inches in front of the wave generator. Over time the sand will begin to migrate shoreward.

3. After several minutes the system will reach a point of equilibrium. As the sand moves toward the shore, how and where does it accumulate? If side view ports are available you can watch the "sand waves" slowly move on shore.

4. Draw or photograph the results after 30 minutes; after 1 hour.

5. Now set the rheostat higher so that the waves are more frequent and have a higher amplitude simulating winter conditions. Again draw or photograph the results after 30 minutes; after 1 hour.

112

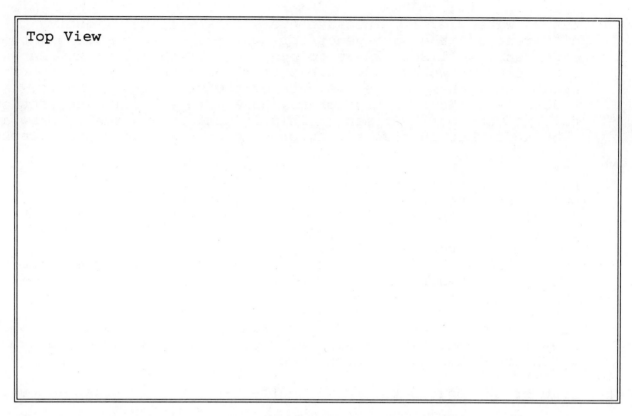

Top View

6. If more than one lab section uses this equipment, either the "summer" or "winter" beach simulation can be run by other sections and the photos or sketches shared. Time-lapse video tape records can be made as well.

Demonstration/Exercise Number 2 - Longshore Drift or the Littoral Current

The seasonal transfer of sand on shore and offshore accounts for the seasonal differences in **beach profiles** (the cross-sectional view onshore-offshore of the beach). Waves also carry sand along the coast. On both the eastern Atlantic coast and the western Pacific coasts of North America, the general movement of the sand along the shore is from north to south. Because most waves are storm generated and because storms are seldom directly in front of the beach, most waves strike the beach at an angle. Thus water and any material being moved by the water will migrate down the shore. This migration of water and material is called the **longshore drift** or **littoral current**. Without such a current, sand being transported by rivers would be deposited at the river's mouth and choke the water flow, yet this seldom occurs. Additionally most sandy beaches occur down coast from the mouths of rivers. In 1967 John Shelton and the film division of Encyclopedia Britannica popularized the notion that the beach could be thought of as a "River of Sand" with production of a film by that title. In their analogy, the "banks" of the river of sand transport were the outer edge of the surf zone and the beach face. We shall see this phenomenon of sand transport in demonstration/exercise 2.

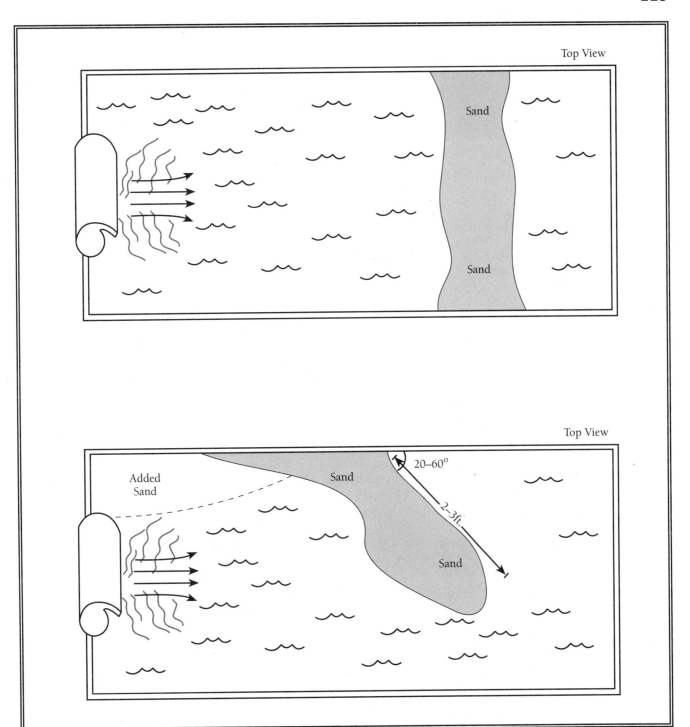

Figure 9-1. Beach face in wave tank. Initial set-up for Demonstration #1.
Figure 9-2. Coast line in wave tank. Initial set-up for Demonstration #2.

Procedures

1. Create, using the sand in the tank bed, a coastline with "beach" angles similar to the magnitude of those observed in the "summer" phase of demonstration #1. Your "coastline" should be at least four inches higher than the water level run at an angle across the wave tank, Figure 9-2. In our tank, our coastline runs for about 3 feet in length.

2. After your features have been created, start the wave generator at a medium level setting.

3. Slowly add the sand provided for you near the generator. This is ordinary lake beach sand, again sieved to remove coarser fragments. (We find our waves won't move the larger pieces; in our scaled-down version they would be car-sized rocks anyway.)

4. Take a photo using the camera available to you or create an accurate sketch as you introduce the sand at the most up-coast location. You now may draw straws to see who the lucky team members are that get to return every other hour of the next 24 hours to take a photo or make a sketch of your demonstration run. After each photo/sketch is taken and developed/drawn, mark on its margin the time taken and place it on the bulletin board in sequence with the others. Be sure to take your photos or draw your sketches so that a clear view of the transported sand is seen and so that each photo/drawing is from nearly the same location and angle. We will analyze our results when we meet in lecture next class time after each group has run the demonstration. If, for whatever reason, any hour photo/drawing is not taken, continue on with the next of the series.

5. A lab report should be written discussing procedures and results. Each team member must sign the report and include a brief discussion about what their role was in the experiment. Also, include a section on what you might tell other groups they should do if they had to do this experiment next week.

Demonstration/Exercise #3: The effects of man-made objects on the beach

Man has had a drastic effect on the shape of beaches and on budgets of the beach sands along both coasts. **Groins** (solid breakwaters extending perpendicular from shore) are common sights along the eastern coast of Florida. Many have been constructed by hotel and other beach front property owners to prevent beach loss or add sand to their beach area. The advantage gained in sand accumulation is very short term. Sand will begin to pile up along the up-current side of the groin until the groin is eventually breached by the shifting sediments. Simultaneously, sand will be eroded from the beach just down-current from the groin, further eroding the delicate beach face. Figure 9-4.

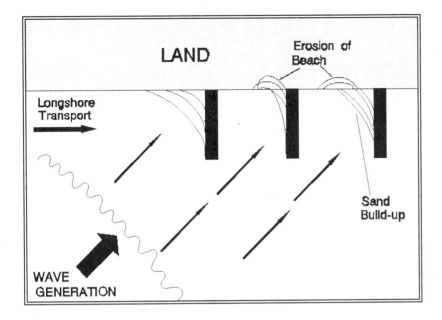

Figure 9-4. The sequence of sand movement when groins are present.

Another example of man's intervention into the system is seen in Santa Barbara, California, where a sand spit is continuously growing at the end of a breakwater built to protect and enhance a boat dock and harbor. Figure 9-5. The sand from the developing spit is removed by continuous dredging and moved only a few hundred meters down stream and redeposited on the beach face so that the sand can again join the longshore transport system. Without constant dredging, the harbor would soon close shut by the sand building up. This process is quite expensive but felt worthwhile by the city government because of the need for such a protected harbor.

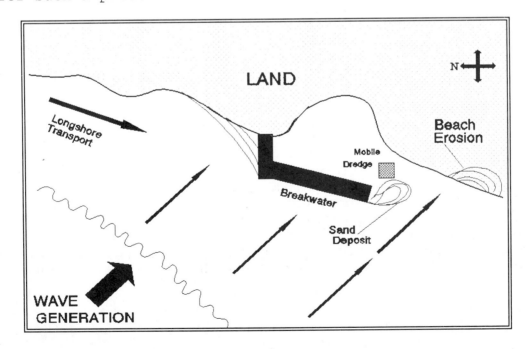

Figure 9-5. The sequence of sand movement at Santa Barbara, CA.

Another example of human intervention is seen at Santa Monica, California, only 80 miles from the previous example. There a breakwater was constructed to provide protection for pleasure craft because no natural harbor existed. After construction of the breakwater, a bulge soon began to appear along the shore behind the structure. The waves which move the sand along the shore were cut off. The reduced flow allowed the sand to deposit and the "river of sand" was robbed of the energy it needed to keep flowing. Again the sand is dredged regularly at a rather large expense to the community. Figure 9-6.

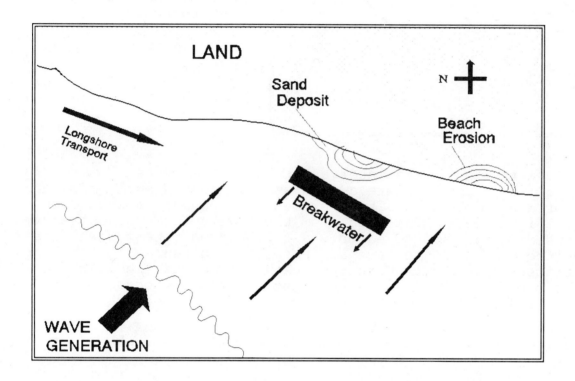

Figure 9-6. The sequence of sand movement at Santa Monica, CA.

1. Create a coastline as in Demonstration/Exercise 2 (the one formed by exercise 2 is fine if it has not been destroyed).

2. Create with a rock(s) or brick(s) of appropriate scale, a series of groins or a breakwater parallel to the coast like those in Figures 9-4 or 9-6 and run the demonstration as you did in exercise number 2. Document your results by photo or drawing after a period sufficient to see the newly created features set up. If time permits, allow this configuration to run for several hours. Record the results. Discuss the results in class.

Questions for Discussion

1. In the introduction to the demonstrations, a discussion of particle size and sand density with respect to creating scale models was offered. What other factors present problems using the wave tank in the lab over actual full-scale runs in nature? Be specific. Can the factors you included be scaled down as well? If not, can the effects be minimized? If so, how? Are there any problems using the wave tank not related to scale? If yes, how can these be minimized?

Answers to question #1

2. If you were a city planner for a coastal area, how might the information or data you just collected influence you with respect to zoning ordinances for beach front property?

Answer to question #2

3. Who should own the sand in transport? Should a land owner be allowed to alter the longshore deposits of sand? Who should decide? Who, if anyone, should be exempt from this law? Explain your answers using arguments you would use pleading your cause before a city or county board.

Answers to question #3

Glossary of New Terms

Longshore current/Littoral current: The migration of water and materials between the shore and the breaker zone.

Groin: Solid breakwaters extending perpendicular from shore.

Further Readings

Film

The Beach: A River of Sand, Encyclopedia Britannica Films, v, Film # 2369 (color), Chicago, 1967.

Printed Material

Pinet, P. *Oceanography: An Introduction to the Planet Oceanus*, West Publishing Company, 1992

Laboratory TEN

Sea Level as a Datum

Sea Level as a Reference Point

While recently climbing a rather small mountain, as far as mountains go, in Tennessee, one of my students turned and in a very out-of-breath voice inquired, "How high are we?" I believe the question was somewhat rhetorical; however, I got out the topographic map of the area, located our position and determined that we were about 5,600 feet above sea level. I relayed the information to my companion who responded in jest, "I hope that is above low tide."

This started a long discussion that centered on the question, "What is meant by **sea level**?" Although this seems like a straight forward question with an apparent straight forward answer, like so many other topics addressed during this course, this too is rather complex when viewed closely. There are several questions to consider about this topic. These include:

- Does sea level in New York have the same meaning as sea level in Papua, New Guinea?

- How permanent of a bench mark is sea level?

- What is the physical evidence that sea level is not a constant?

- What can cause sea level to shift?

- Are sea-level fluctuations normal?

- How often do measurable fluctuations occur?

- Are these fluctuations rhythmic, cyclic or random?

- Does it really matter if sea level is not constant?

Addressing the last question first, I would have to say that it does matter from the perspective that some worldwide, uniform reference plane is useful, if for no other reason than to answer my hiking companion's questions.

Water level appears, at first, to be the ideal medium to use for a reference plane. Most of us have been taught or observed two principles of water that we should never forget:

- Water always tries to seek its own level.

- Water will always take the path of least resistance to achieve the above.

Using these two axioms, let us see if we can create an imaginary world where water covering the planet could form a truly uniform datum. Let's imagine a perfectly homogeneous, spheroidal earth with no spin, no land masses interrupting a continual, watery envelope, and no neighboring celestial bodies. Left to gravity alone, the water at the surface would cover the earth at an uniform level. Before continuing, create a mental list of ways in which the earth does not fit this simple, imaginary world.

We do not live on such an imaginary planet. The earth is not spherical; some places are closer to the center of the planet than others. The distance of any location on the surface from the center creates, to a small degree, greater or lesser forces of gravity acting on that location. The earth is made of materials of varying densities as we saw in an earlier laboratory, and not composed of a homogeneous mixture of rock. Because density effects gravity, this too can affect the level of gravitational attraction. The levels of the sea are also affected on a daily basis by the attractions and position of both the sun and the moon in the form of tides. And, of course, there are land masses that interrupt the flow of water from circumnavigating the world.

As the earth spins and the sun shines more directly on some locations than others and as land and sea heat and cool at different rates, high and low pressure systems are developed. As the atmosphere attempts to return to a state of equilibrium, winds are created. The results are broad bands of nearly unidirectional winds at various latitudes that, with the rotation of the earth, combine to stack water along the eastern coasts of continents and pull water away from western coasts. The net effect is for higher sea levels to develop on one side of a continent preferential to the other. Over a large landmass like North America, this difference in "sea level" can be averaged out on topographic maps with no noticeable effect. In a narrow land bridge like Central America, however, this difference can be recognized. Nearly a 20 foot difference in the levels of the Atlantic and Pacific Oceans exist on either side of Panama. This is not even the greatest standing sea-level difference. Between Papua,

New Guinea, and just south of Sri Lanka, there is over 192 meters of difference in distance from the center of the earth to the sea's surface. Yet each site is considered sea level, or is it?

For mapping purposes the United States Geological Survey (USGS), in 1929, using several stations on the east and west coast of the United States, determined a standard, if artificial, level for the sea on topographic maps of the United States. What this means is that, to keep the problems mentioned above from confusing the issue when mapping, a uniform datum was agreed upon and has been used ever since. A recent problem has been noted. Over the past seventy years, sea level has risen substantially (over one foot) in some areas and has caused cartographers much consternation, so much so that a new plane of reference has been agreed upon for mapping sea level. This new plane will be used on maps being created currently and as updated versions of older maps are made. It is assumed that other countries will soon follow the USGS's lead.

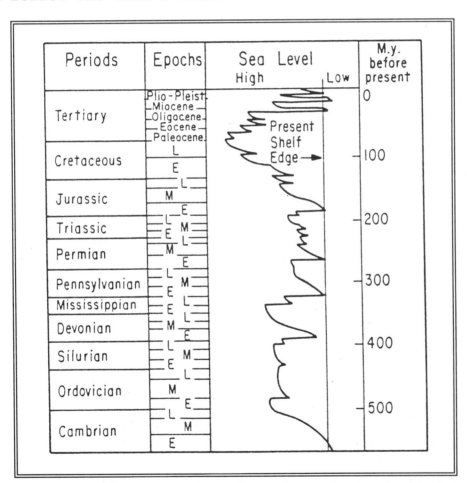

Figure 10-1. Inferred sea-level curve for Phanerozoic time. (From R. Vail, R.M. Mitchum Jr. and S. Thompson III. *Seismic Stratigraphy: Applications to Hydrocarbon Exploration*, 1978, **reprinted by permission.**)

In 1978, researchers at Exon Production Research lead by Vail,

Mitchum and Thompson published a report showing inferred sea-level fluctuations through Phanerozoic time, roughly from 550 million years ago to the present (Figure 10-1). Although this report only indicated relative changes in sea level, it was one of the first attempts to discriminate apparent sea-level shifts worldwide from the apparent shifts that may be more local in nature. Since the publication of the "Vail Curve", several others have tried to refine the details using more recent data. V. Gornits and others, in 1982, published tide gauge data collected since 1880 that show an overall rise in sea level of about 4 inches over the last 100 years (Figure 10-2). Although that does not seem like a very large amount, we will see, in exercise two, that the results can be quite staggering. Figure 10-3 basically bridges the time scales of the data shown in Figures 10-1 and 10-2. This graph shows the changes in sea level during the past thousand years. This is the interval of the last great North American glacial ice advance and retreat.

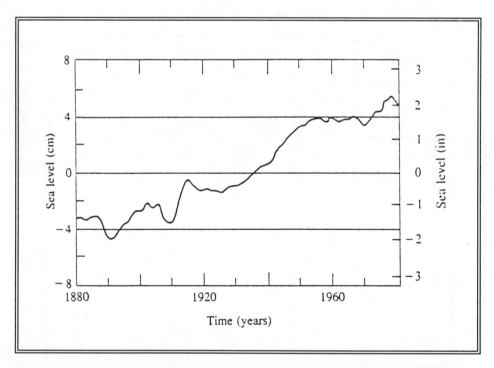

Figure 10-2. Sea-level change from 1880 to 1980. (Reprinted by permission of Macmillan Publishing Company from *Essentials of Oceanography* by Harold V. Thurman. Copyright 1990 by Macmillan Publishing Company.)

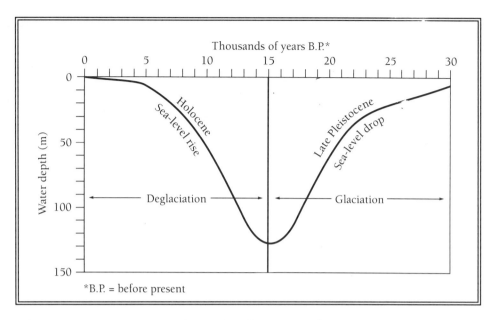

Figure 10-3. Sea-level fluctuations over the past 30 thousand years. (Adapted from K.O. Emery. The Continental Shelves. *Scientific American*. 221:3, 1969.)

Exercises

1. **Stranded Beaches**
    This exercise will again use the wave tank. About half way down from the wave generator, create a peninsula extending from either side of the tank. The peninsula should extend about 3/4 across the width of the tank. This demonstration will work best if your landform extends about 4-6 inches above the water's surface. Water depth should be at least 7-8 inches to see the results of this experiment most readily. (Figure 10-4.)

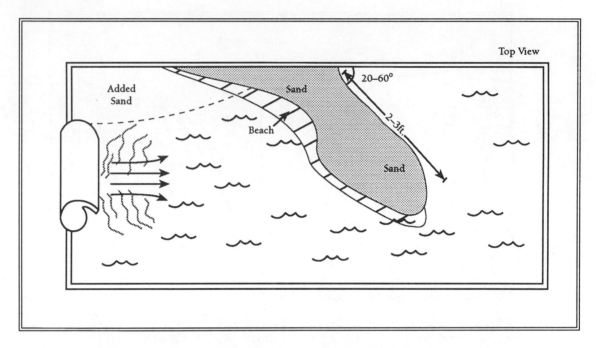

Figure 10-4. Wave tank set up for exercise #1.

Allow the wave generator to run for about 10-20 minutes or until a well-developed beach forms along the seaward face of the peninsula. Then drain approximately 1-2 inches from the tank while the wave generator is still running. Allow a new beach face to develop at this new "lower stand" of sea level. After the new beach forms, repeat the procedure once more so that three well- developed beaches are present.

Examine the profile (cross-sectional view) of the shoreline from some point on the highlands to a point a few inches offshore. To best see the features, your profile observation point should be perpendicular to the coast. A piece of notebook-sized, white poster board slid carefully into the sand perpendicular to the shoreline without destroying the beach structures, may help you see the structure more readily. Notice the flat plains made as the beaches developed during times the water level stayed relatively constant. The plains alternate with the sloping beach faces which formed as the level dropped. You saw a similar profile on the map of southern California in the exercise on coastal classification. These features are stranded beaches and they occur when either sea level falls or the land rises in an interrupted fashion. The stranded beaches are often used to help date sea-level fluctuations. They are especially useful if the materials associated with them, like fossil shell beds or wood, are usable as time indexes or markers.

2. "Swamping" Florida
This exercise is designed to give you an accurate depiction of what will happen if sea level continues to rise at the same rate as Gornits and others have said it currently is (Figure 10-2). In a recent article Doyle (1991) has noted that the "Shorelines

in Florida have been eroding over the last 40 years because of massive development and the natural evolution of coastal erosion ... the burning of fossil fuels, beginning about 1700, has led to the green house effect and the melting of the glaciers ... A slight increase in sea water has a tremendous impact on shore retraction. A centimeter increase in sea water causes a shoreline to retract 500 times as much." Using the topographic map of Florida provided, shade in all areas where land is currently below an elevation of 50 feet. The landform left unshaded is greater than that which will remain in 10,000 years if rates measured over the last 100 years are maintained. The paucity of Florida left as dry land may surprise you. It could certainly affect land speculation! Of course, this rate is extremely speculative and improbable, we hope.

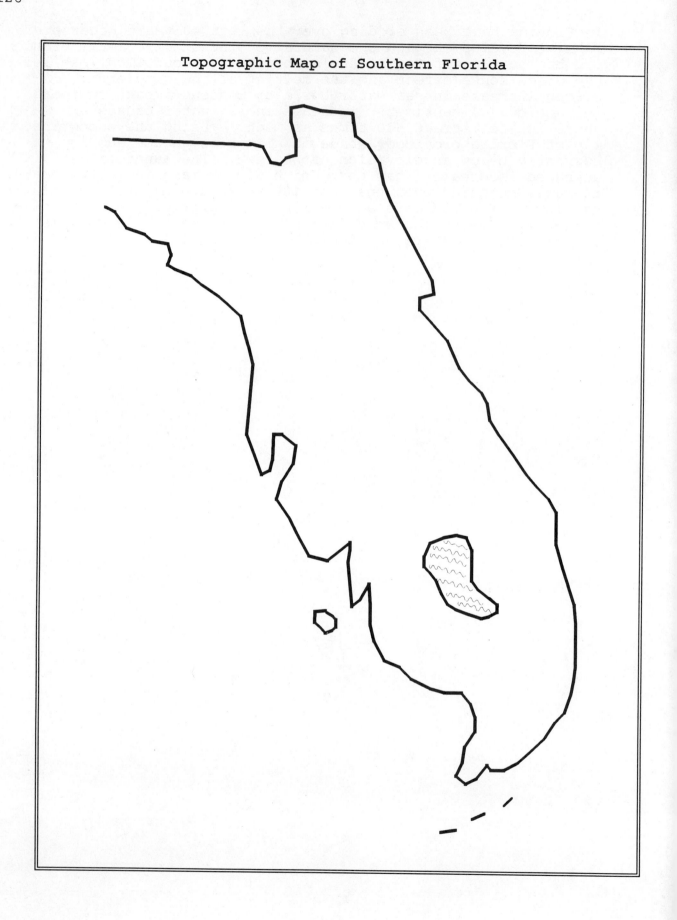

a. Where else might this modest rise affect great groups of people?

b. What do you believe is the smallest amount of sea-level rise that could be sustained in the United States without a great disruption to the citizenry?

c. What criteria did you use to base your judgement? Justify your response.

Answers to question #2
a.
b.
c.

3. **Eustatic Changes in Sea Levels**
Of course there are many factors that affect the level of the sea. A few of the most well documented are listed below. After each, discuss how this factor affects sea level and the magnitude of the effects due to that factor. That is, are the effects seen locally, regionally, or globally? Is the shift in sea level daily, yearly, century-long, or measured in time greater than tens of thousands of years.

Factors affecting sea-level fluctuations
Sun/Moon/Earth interaction:
Hurricanes:
Active Tectonism, i.e. increased rate of sea floor spreading:
(Continued on next page)

Isostasy:
Continental Glaciation:
Rapid Continental Erosion:

4. **Sea-level Change II**

   Of course, the ratio of land/sea surface area is greatly affected by changes in sea level.

   a. What other physical changes occur near the land/sea interface like on the continental shelves?

   b. What effect do these changes have on the **niche** and **habitats** (how organisms "make their living" and where they live) of organisms living near shore? For example, if continental glaciation began rapid advances again, the water for ice production would ultimately come from the oceans. As ocean levels declined, the amount of sea floor within the **photic zone** (the depth of light penetration from the surface) would shrink drastically. Some continental shelf areas might be completely exposed. Obviously, this would be disastrous for immobile marine organisms.

   c. What other conditions would result that could greatly affect biological **diversity** and **abundance** in the fertile continental margins?

Answers to question #4
a.
b.
c.

Glossary of New Terms

Abundance: Total number within types or kinds.

Diversity: Total number of types or kinds of organisms.

Habitat: Where an organism lives.

Niche: How an organism "makes its living."

Photic Zone: Depth in which sunlight can penetrate the seas.

Further Readings

Carr, A.P. The ever-changing sea level. *Sea Frontiers*, 202, 77-83, 1974.

Dole, L. Short Note. *Geotimes*. April, 1991.

Fairbridge, R.W. The changing level of the sea. *Scientific American*. 202:5, 70-79, 1960.

Feazel, C. The rise and fall of Neptune's kingdom. *Sea Frontiers*. 33:2, 4-11, 1987.

Vail, R., R.M. Mitchum Jr. and S. Thompson III. *Seismic Stratigraphy: Applications to Hydrocarbon Exploration* ed. C.E. Payton. AAPG, Memoir 26, 1978.

All the rivers run into the sea; yet the sea is not full.

The Bible, Ecclesiastes 1:7

Laboratory ELEVEN

Deltas

Deltas and Delta Formation

When moving water carrying sediments drains into an area of standing water the sediment load is dumped. The resulting deposit is known as a **delta**. The greatest percentage of sediment that ultimately forms sedimentary rock is the material deposited in deltas and along shorelines. Nearly every stream system produces some form of delta at its mouth, even those on sidewalks draining to puddles. The shape of the delta formed depends on several factors including:

- The amount of sediment being deposited

- The sediment budget of the long shore current

- The geometry at the mouth of the stream

- The erosive force of the sea's waves and currents.

The variety of shapes that deltas may form is the basis for their classification. When the amount of material entering the coastal system is balanced with the erosive energy of the sea, as in the case of the Nile River, a broad triangular plain resembling the Greek letter Δ (Delta, hence the name) will form. When the sediment load is much greater than the ability of the sea to remove it from the area, as at the mouth of the Mississippi River, a broad platform of sediment with long finger-like distributary channels extending far from shoreline will form. These deposits are known as bird foot deltas. Where the energy level is higher than the depositional rate of the sediment, as at the mouths of the Amazon and Niger Rivers, where the rivers empty into bodies of water with great amounts of wave

and current energy present, few deltas will form at all. In a very
few cases, not enough sediment is carried to form a delta. An example
of such a place is the mouth of the St. Lawrence River. See Figure
11-1 for examples of the various delta geometries described above.

   Deltas can be quite complex depositional systems, and recognizing
ancient deltas in the rock record can often prove difficult. Once
again, it is not necessary for you to completely understand the
intricacies of delta formation to see how various sets of
environmental conditions affect their growth or their geometry.

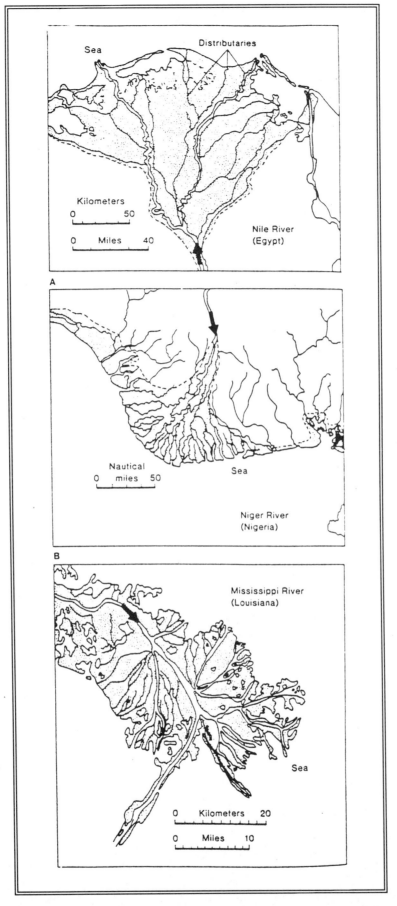

Figure 11-1. Examples of delta styles (From Charles C. Plummer and David McGeary, *Physical Geology* 5th ed. Copyright(c) 1991, Wm C. Brown Publishers, Dubuque, Iowa. All rights reserved. Reprinted with permission.

Laboratory Discussion

Today we will return to the wave tank or stream table and attempt to recreate a few of the delta types described in the introductory material. You will note that the wave tank looks much different from the last time we used it. Now it has had a great deal of sand and soil sediments added to it. Note that the land mass created has been pushed to the end opposite the fan (wave generating device), and a landscape about 3-5 inches deep has been formed. The end nearest the wave generator remains sediment free and has the standing water (1-2 inches deep), needed to form our deltas. We have found that a slight 1-1½ inch barrier set along the bottom across the entire width of the tank keeps the landform from slumping into the pool and initially keeps the two systems apart.

In nature, there is no need to separate the systems as they are literally extensions of one another. However, for us to watch the development of the deltas we will form, it is best not to have sediment depositing on top of pre-deposited sediment. The demonstration/exercise will work best if there is a slight grade downward on the land's surface toward the water. To get the meander system started you may need to create a channel with your finger at the top of the surface near the water source. A hose connected to a water source is mounted by a ring stand slightly above the land at the end of the tank. This setup may seem familiar to some of you, as this stream table configuration is a common one for many demonstrations in classes dealing with earth or soil sciences. Figure 11-2 can be used as a reference to show how the equipment should appear at the onset of the exercise. As equipment may vary from institution to institution, there may be slight differences in their appearance when set up for this demonstration/exercise.

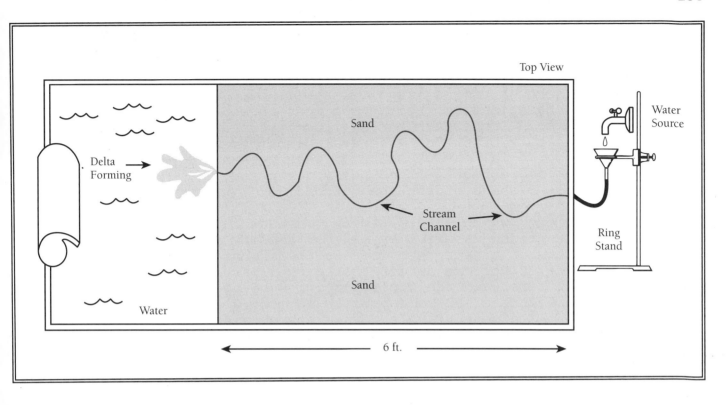

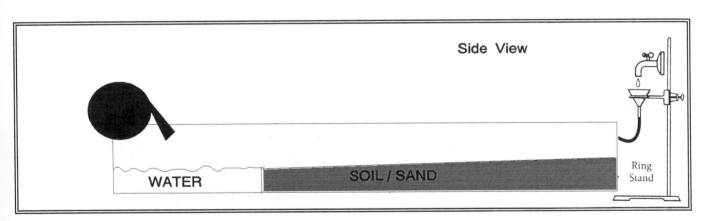

Figure 11-2. Lab set-up for stream table and delta formation.
a. Top view. b. Side view.

Demonstrations/Exercises

Demonstration/Exercise #1

Turn on the water such that a slow, steady stream begins to trickle from the suspended hose. The first bit of water will probably absorb into the sediment, but soon a small stream channel will form and begin to meander down slope to the standing water. If you look closely, you will see that sediment is being carried along with the moving water. The sediment will deposit when the stream finally completes its course to the standing water. The first run of this demonstration will be done with no wave or current energy added to the system. Over enough time a large delta will form and you may see distributary channels form, become sediment choked, abandoned and perhaps reformed. Since each run of a demonstration like this is unique, we can only offer you a general description of how the results will look. Your trial may be quite different than the ones we describe; that is perfectly fine. As you run your demonstration, consider what variables effect the outcome.

One parameter which should remain constant in each run is the sorting of the eroded material by size or density as the velocity of the stream slows to a halt when it hits the standing water. Watch closely and see if you can detect the coarser particles settling out of the current first and the finer particles being carried further out to "sea". You can see the same process in mini-delta formation after every rain when sediment travels down-slope on sidewalks or in parking lots and deposits in standing puddles. I frequently surprise my class by taking a spontaneous field trip outside our class building during or after a good rain.

Once you are satisfied that you understand the development of a simple delta, **turn off the water** and proceed to the next exercise.

Demonstration/Exercise 2

We will now begin to add variables to each run of the experiment. Push back the material that formed the delta in the first exercise so that you are back to your initial set-up. Take the small barrier used to hold the shoreline in place and set it at a slight angle to the walls of the tank (see Figure 8-3) so that the shoreline formed is not perpendicular to the sides and angles in front of the wave generator, stack the sediment behind the moved barrier. When the wave generator is turned on, the waves should not strike the beach "head on", but at an angle such that a long shore current will develop. To do this best, you may need to contrive the stream channel so that it deposits and forms a delta somewhere up the long shore current (about two thirds of the way up shore seems to work well). This will keep the effects of waves rebounding off the wall at the end of the shoreline to a minimum and prevent them from altering the shape of the delta.

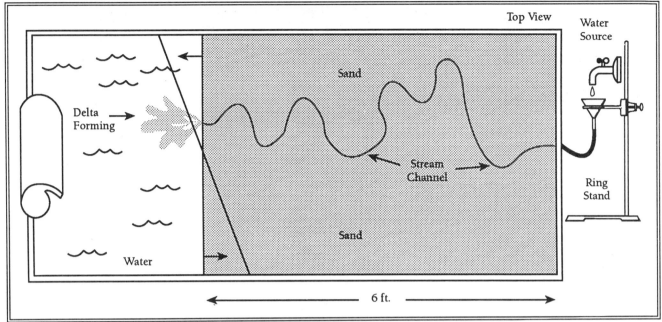

Figure 11-3. Set-up for Exercise #2.

If the wave tank/stream table you are working with has the capacity to add current to the system, this may also be applied to the delta being formed and its effect noted.

Draw, as accurately as you can (or photograph with an instamatic camera), the results of each of your trials. You may wish to stop the water flowing from the hose for a few minutes at a time to allow you to sketch the system at various stages of development. You can restart the exercise simply by turning the water back on. If you let the experiment continue for several hours, you will probably see another phenomena of running water. The erosion of the stream channel will become lateral instead of a vertical down-cutting as the water in the stream reaches its **base level**, the level of the water in the catch basin.

Glossary of New Terms

Base level: The level of the standing water in which streams deposit.

Delta: Deposit formed at mouth of a stream.

Further Readings

Morgan, J.P. Deltas, a resume. *Journal Geological Education.* 18:107-17, 1970.

Morgan J.P., and R.H. Shaver, ed. *Deltaic Sedimentation.* Tulsa, OK: Society of Economic Paleontologists and Mineralogists. Special Publication No. 15, 1970.

Pinet, P. *Oceanography: An Introduction to the Planet Oceanus,* West Publishing Company, 1992

Laboratory TWELVE

Beach Sand

Beach Sand Diversity

As discussed in the introduction for the lab sediments and sedimentary rocks, sand is a size fraction ranging from 1/16 mm for very fine sand to 2 mm for very coarse sand. Notice that no material or composition is implied in this definition. Although many tourists would disagree, to a geologist and a physical oceanographer, sand is sand whether it is black, white, coral, yellow, gray, red, brown, pink or green (all colors that can naturally appear in sands, as you will soon see).

There are several important factors for oceanographers to address when analyzing sand. These include:

o   Why does the sand collect where it does?

o   Where does the sand originate?

o   Where is it going?

o   Does the sand's presence alter any other environmental parameters?

o   Will the sand always be where it currently is?

You started to address these questions and more during the last several labs, working with the wave table. One observation many have made when describing the results of the sand transport lab is that the deposits on the beach face could vary dramatically within only a very short distance. Sometimes in nature, as little as ten centimeters can separate sands that are coarse from those that are very fine. This is frequently seen in the sand deposited around obstructions on the beach. The objects cause eddy currents to form around the obstructions. These eddy currents in turn deposit various sized sand around the object as the current flow ebbs and wanes. As we saw in an earlier lab, on a bit larger scale, different sized sand can frequently be collected from different water depths in an on-shore/off-shore sequence.

Size is not the only criteria of sorting by the currents on sandy beaches. Mineral density also plays an important roll as does shape, to a lesser degree. This sorting was seen in the sand-filled jars that were shaken and allowed to settle earlier and will be seen readily in today's lab.

Exercises

We will be using binocular microscopes, much like ones used in biology courses for dissection. They can magnify up to 30 times. For the best results in describing the sand, we have found that starting at around a ten times magnification and working up to the maximum works best.

Like the last few labs, the answers you give will be based on your (and perhaps combined with your lab partners') experiences, observations, and your deductive reasoning power. You need not consult other sources for the answers to the exercises. As a matter of fact, it is preferred that you do not. Try to "think out" the solutions.

1. **Microscopic Inspection of Beach Sand**

    Prepare a small sample of the beach sand out of the containers from "known" locations for microscopic inspection. The locations are labeled on the containers. A small pinch of less than a dime's surface area is enough of the sample to use for this procedure. Before viewing under the scope, try to describe in one or two sentences what the grains look like to the unaided eye. Color, size, texture (sharp, rounded, angular, blocky, etc.) and , if you can, any minerals that are present are among the parameters you should note. **Don't strain your eyes!**

Description with unaided eye

Now look at the sand under magnification. Redescribe the sand.

Description under 10-20x magnification

Is there a difference in your observations? Why or why not? Compare your descriptions with those posted in the lab that were prepared for the samples at an earlier date. The purpose of this exercise is to show you the difference of scale that observation can have. Take another look to see if you can see the reasons for differences in yours and the one done for you. Return your sample (if it has not been contaminated) to the sample container so that it may be reused in later labs.

2. **Comparing Beach Sand to Patent Rock Material**

Set up for you in lab is an example of sand from along the northern coast of California and an example of the original rock parent collected in the mountains inland, from which it came. How are the rock and sand compositions different? Why are some of the minerals present in the rocks not in the sand? Refer to lab chapter three if you need assistance. Design an experiment that might give you a scale of the relative hardness or resistance to erosion of the minerals that make up the **pegmatitic** (very coarse crystalline) granite parent rock of this sand. Yes, you can do this just by observing what you know exists in the geographic area where both the sand and the rock samples were collected. Referring to a map of California may assist you.

Differences between parent and daughter compositions

Reasons for differences

Experiment design

3. **Sand Gradients Along a Beach**
   The samples marked 3A, 3B, 3C, 3D were taken from a beach in southern Florida two feet from each other with 3A being above the slack high tide water level, 3B at the water level, and 3C and 3D in the water going away from the shore. Waves were minimal during collection.

   a. What different interpretations could be formed about this beach if only one sample had been seen?

   b. Which sample is most diagnostic?

   c. Would this be true for all beaches? Why/why not?

   d. Discuss how your image of the beach changed as you worked your way up or down the series of samples.

Answers to exercise #3
a.
b.
c.
d.

4. **Analyzing Sands of Unknown Origin**
   Now choose at least two sands from the "unknown" group. Describe them in the same manner. Using the combined resources of those working with you, try to determine from where the sand may have originated (what kind of rock). Be sure to include why you believe this. Take a guess, if you'd like, as to where in the world geographically it may have come from. Why are exercises like these important to an oceanographer's training? When might he/she use this type of information?

Description of sand samples
Sand sample #1 -
Origin -
Sand sample #2 -
Origin -

Oceanographer's use of this information

5. **Optional**

   A collection of sand from various beach locations around the world has been in the making for quite some time. View as many under the microscope as you have time. It may surprise you how different in composition they are.

   You too can be a celebrated sand contributor. To help us expand the sand collection for lab use, any time you go to a naturally formed beach, collect a bag full of sand and send it to your institution. Your name will be forever enshrined on the sample jar. Be sure to include all pertinent information such as exact location, type of environment (ocean beach, lake shore, river bed, dune, etc.), date, any special observations, and the name of the collector.

GLOSSARY OF NEW TERMS

Pegmatitic: Very coarse crystalline igneous rock.

FURTHER READINGS

Kaufman, W., and O. Pilkey. *The Beaches are Moving*. Garden City, New York: Anchor Press/Doubleday, 1979.

Kennett, J.P. *Marine Geology*. Englewood Cliffs, N.J.: Prentice-Hall, 1982.

Pinet, P. *Oceanography: An Introduction to the Planet Oceanus*, West Publishing Company, 1992

Siever, R. *Sand*. New York: Scientific American Library, 1988.

Now I hear the sea sounds about me; the night high tide is rising, swirling with a confused rush against the rocks below... Once this rocky coast beneath me was a plain of sand; then the sea rose and formed a new shore line. And again in some shadowy future the surf will have ground these rocks to sand and returned the coast to its earlier state... the earth becoming as fluid as the sea...

Racheal Carson,
The Sense of Wonder

Laboratory THIRTEEN

Marine Microfossils

Foraminifera, Sponges, Bryozoans, Etc.

In exercises nine and twelve we viewed several types of beach sand, yet they had in common the fact that they were primarily composed of rock fragments derived by the erosion of larger rock pieces. Today you will view more sediment, but this sand is primarily composed of materials of organic origin. In most cases these grains are principally $CaCO_3$, or the mineral calcite, although you will view some composed of SiO_2, or the mineral quartz.

Because of the ease of access, most of the microfossils viewed in this lab will have been collected along beaches. Beaches are not the most important deposits of this type of material. Much of the deep sea floor is covered with deposits primarily composed of tests (shells/skeletons) of these organisms. Figure 13-1 shows a generalized map of these deposits. One can see on this diagram that a high percentage of the sea floor is made up of either calcareous or siliceous sediments. these are primarily composed of microfossils. The calcareous ones ($CaCO_2$) litter the moderate depths and mask the presence of the much smaller and less common siliceous organisms. However, at the deepest depths siliceous sediment dominates and no calcareous material is present. Why is this? Both kinds of organisms are free floating and live in the shallow depths of the oceans world-wide. Why is there a difference in where they deposit? The answer lies in the chemical structures of calcite and quartz. Under the great pressures of the deep sea, calcite will go into solution and not precipitate. The depth that this occurs varies with temperature. The depth is shallower at the equator and deepens at the poles. It is still a depth greater than 3 kilometers even at the equator. This level is known as the Calcium Carbonate Compensation depth (CCD).

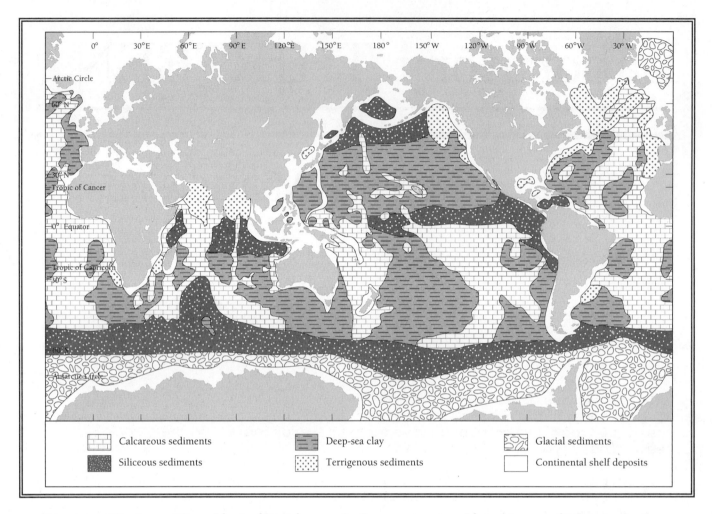

Figure 13-1. The distribution of deep-sea sediments. (Adapted from T.A. Davies and D.S. Gorsline. Oceanic Sediments and Sedimentary Processes. *Chemical Oceanography*. J.P. Riley and R. Chester (editors), V5. Orlando, FL: Academic Press, 1976.

So as the organisms living near the surface die, they begin a descent to a sea floor resting place. The descent may take decades as the tiny shells are buffeted about by even the smallest of currents. If the sea floor is above the CCD, then all of the materials can be preserved and the larger more abundant calcareous material dominates. If the sea floor is below the CCD, then the calcareous material dissolves into the water and only the siliceous material remains to form the bottom deposits. By now it is probably obvious that the calcareous sediments accumulate much quicker and are most likely thicker. This is the case most often.

You will see when examining the grains of sands under the microscope, magnified 10-30X, that many are actually whole skeletons or shell parts of various marine organisms. These shells or **tests** are from a group of nearly microscopic organisms known as Foraminiferida, a subclass of Sarcodina. These unicellular animals range in age from the Cambrian (over 500 million years ago) to the present. These organisms are commonly called "forams". There are over 30,000 recognized modern species of forams. You will not be asked to identify any of these taxonomically. Your exercise will be to pick and mount as many different kinds of organisms or organism fragments as possible and identify by shape as many forms as you can from the two sand sources in the lab.

To assist in identifying the specimens you've mounted, a chart illustrating common forms found in the sand samples is included (Figures 13-2 and 13-3). Forams are separated into various subgroups based on the construction of their external tests or covering and the material of which it is made. The word shell here is not always accurate and should be avoided. The following is a list of typical test materials for forams.

TESTS NOT MADE UP OF $CaCO_3$:

Agglutinated:
These tests are external coverings composed of loosely cemented grains and foreign matter picked up by the organism from the sea floor. This group can be identified by the irregular appearance of the grains under the microscope. They are not common, but can be seen in the samples in today's lab.

TESTS MADE UP OF $CaCO_3$:

Microgranular:
Example: Fussilinids. These tests look like grainy limestone shells with coarse texture and are usually dark gray in color. They look like wheat in size and shape. When cut across the long axis, one can usually see that they grow by adding material to the outside and look in cross-section like a crescent roll.

Hyaline:
These tests look shiny and glassy and are translucent to transparent, **not white**. They often look and are very delicate. Hyaline tests come in many shapes.

Porcellaneous:
These tests look like the broken edges of fine china and often appear white like chalk. They are most often a bit larger than the hyaline forms. Both the hyaline and porcellaneous forms can be quite complex and varied.

Forams are also subdivided by the shape of the test. Nearly all have one or more enclosed chambers within the test. Some appear quite simple, such as those which are bottle shaped. Others have complex coiled or stacked chambers. A variety of the shapes you may find are shown in Figure 13-2. Refer to it often as some shapes such as the biserial and triserial can be very confusing and may appear, at first, to be the same. When identifying the forams, for our purposes, it will be sufficient to name the material of the test and its general shape. One readily identified foram is the globular form, *Globegerina*, seen commonly in the Timm's Point sample. This foram is a large contributor to the calcareous oozes on the ocean floor discussed earlier. Figure 13-1.

There are other organic materials present besides forams in the sand for today's lab. Several small (< 3mm) glass rods in a variety of shapes, from simple single rods to complex forms that look like jacks from the children's ball game, have been seen. Some can even look like glass, elongated, bumpy wheat grains and can range in color from purple to yellow to clear. These "glass" structures are made of quartz (SiO_2) and are formed by sponges (Phylum: Porifera). The rods are called spicules. Not all sponges make this form of spicule, but those that do are fairly common as sediment contributors in some geographic areas.

There are other organisms present as well. Mollusc fragments and the protoconchs of gastropods (juvenile snails mostly) are frequently encountered. These can be recognized from forams because they are unchambered and generally a bit larger in size. Fragments obviously will be difficult to identify. Echinoid spines are common as well. These look to some like corn cobs and come in a variety of colors with yellow, white and purple dominating. Fish teeth, vertebrate, fish scales and fecal pellets are also found occasionally. The less frequently seen items and examples of the common forams and sponge spicules can be seen in the reference slide set provided in the front of the room. As you will see, some of the drawings in Figures 13-2 and 13-3, while accurate, cannot do some of the test forms justice. The reference set is provided to augment the line drawings.

Exercises

1. **Pleistocene Marine Microfossils**
   The sand marked "Timm's Point" is from Northern California, just north of where the pegmatite and sand problem specimens were procured in an earlier lab exercise. This material was collected along a beach of sandstone of Pleistocene Age (about 50-70 thousand years old). The "fossils" are mostly foraminifera - single celled organisms that often live floating on the water and sink to the bottom after they die. Although some live in the bottom sediment, they all become incorporated into the sand sediment matrix. Notice that this sand has grains of both organic and inorganic material. Estimate the percentage of each (the organics **vs.** inorganics). Prepare a mounting slide by smearing a water diluted solution of white glue over the entire black grid surface of the mounting slide provided. Let the slide dry completely. Be certain to write your name on your slide. After preparing a mounting slide, under the microscope choose at least 15 **different** kinds of organic grains. Try to use whole ones. Mount the grains on the grid boxes marked 1-15 on the slide you prepared by wetting a **00000** sized paint brush in water and removing excess water along top rim of beaker. Then pick individual grains by touching them with the fine damp bristles

(the grains will stick to the brush). Then "paint" the brush across the grid area on which you wish the specimen to appear. The glue will partially dissolve and allow the fossil to stick to the square. Practice makes perfect; don't get frustrated! Prepare the identification key provided for you. Spaces 16-25 are for your use - have fun!

2. **Recent Marine Microfossils**
The sand marked "Little Saddleback Cay" is a carbonate sand from near Andros Island, Bahamas. The sand has no rock fragments included in it; it is made up 100% of microscopic sized and larger organisms or fragments including mollusc shells, corals and many calcareous algae. In the front of the lab, displayed for your use, are examples of several identified calcareous algae from the area that create hard parts for themselves by secreting $CaCO_3$ from the sea water. Most of the algal material will be difficult to identify beyond the fact that it is algae. Why do you suppose this is true when the rest of the sand material of the same composition is so well preserved?

When these sediments accumulate, are put under pressure (the weight of overlying material is enough), and calcium rich water percolates through the sediments, they will fuse and form not sandstone, like the rock sediments viewed before, but a grainy limestone known as **coquina**. As the pressure increases and more water percolates through the pores, crystals of $CaCO_3$ enlarge, fill the void spaces, and the grain boundaries become nearly impossible to differentiate. The rock formed then becomes a massive limestone. Example of coquina and massive crystalline limestone are displayed for you to view in the lab.

Collect organic grains (15) from the Little Saddleback Cay sediments and mount them in spaces 26-40. Again, identify them as best as you can by composition and shape. Refer to the diagrams and the lab displays.

3. **Oolites**
The sample labelled "Joulter's Cay", contains another form of carbonate grain, but these are **not** organic in origin. They are **oolites** and can be thought of as carbonate snowballs. As the grain in the core is rolled around along the carbonate beach, it adds layer after layer of aragonite (a form of $CaCO_3$). When cut in half, oolites look like concentric circles or bull's eyes. Prepared slides of dissected ooids are available for view in the front of the lab. Please be careful with these "thin sections"; the glass slides break easily and are not easily replaced. Mount 5 ooids in spaces 41-46. Draw in the space provided a cross sectional view of an ooid in thin section. Ooids have hard-packed cores much like large snowballs used to make snow people. Most ooids form around a fragment of shell or other broken piece of organic material. Often the core of the ooid is still identifiable, at least to the phylum in which the fragment maker belongs. Usually a cross sectional view must be seen magnified 20-30X to see the core clearly. Can you identify the core origin

in any of the ooids on the slide? If so, what do you believe they are?

Ooid Cross-Section

Scale

4. Radiolarians

Although not normally found in beach sands, another type of microfossil produced by another single celled organism is also important in marine sediment. These organisms are Radiolarians and their tests are normally composed of silica in the form of quartz or opal. Radiolarians form a subclass of the Sarcodina and are the principle constituents of siliceous deep see oozes. Figure 13-1. As a rule these are about one order of magnitude smaller than the forams and, as such, would be extremely difficult to pick and mount. A slide of various radiolarians is prepared for you to view. Draw two of the forms you see. Try to include as much of the detail as you can.

Radiolarian

Scale

Radiolarian

Scale

Micromount Key

Timm's Point California, (Pleistocene) Beach Sand		
% organic _____		% inorganic _____
1 _____	10 _____	18 _____
2 _____	11 _____	19 _____
3 _____	12 _____	20 _____
4 _____	13 _____	21 _____
5 _____	14 _____	22 _____
6 _____	15 _____	23 _____
7 _____	16 _____	24 _____
8 _____	17 _____	25 _____
9 _____		

Little Saddleback Cay, Andros Island, Bahamas (Recent) Beach Sand		
26 _____	31 _____	36 _____
27 _____	32 _____	37 _____
28 _____	33 _____	38 _____
29 _____	34 _____	39 _____
30 _____	35 _____	40 _____

Ooids from Joulter's Cay, Andros Island, Bahamas	
41 <u>Ooid</u> _____	44 <u>Ooid</u> _____
42 <u>Ooid</u> _____	45 <u>Ooid</u> _____
43 <u>Ooid</u> _____	

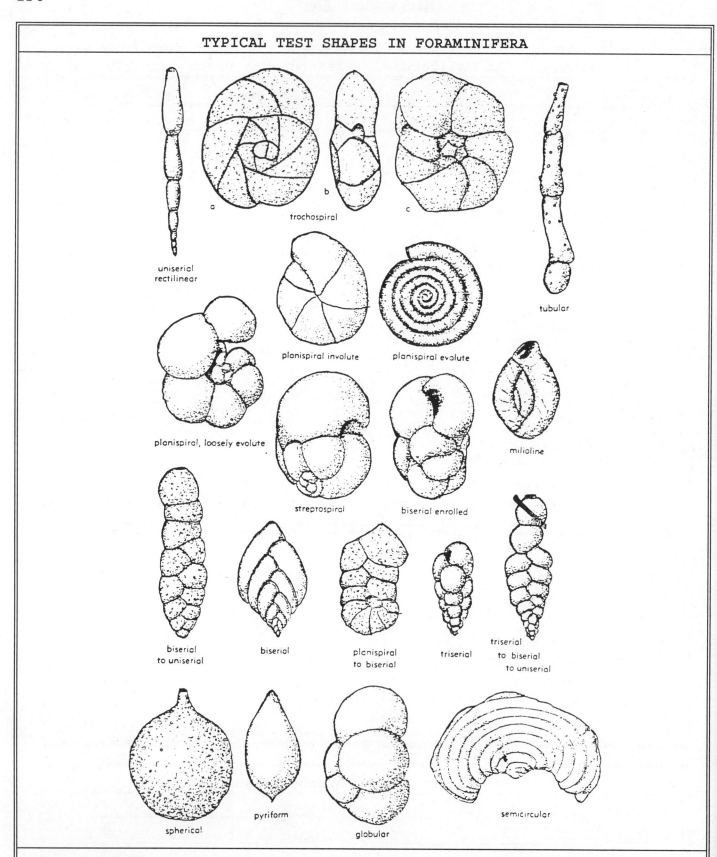

Figure 13-2. (From *Treatise on Invertebrate Paleontology*, courtesy of The Geological Society of America and University of Kansas.)

Typical Forms of Siliceous Sponge Spicules

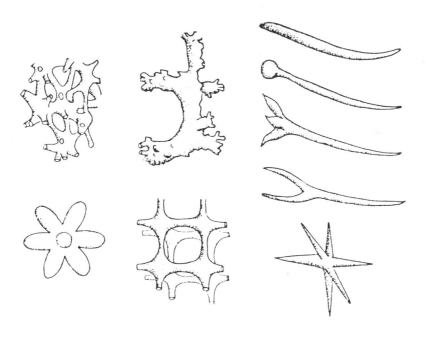

Typical Representative Radiolarian Shapes

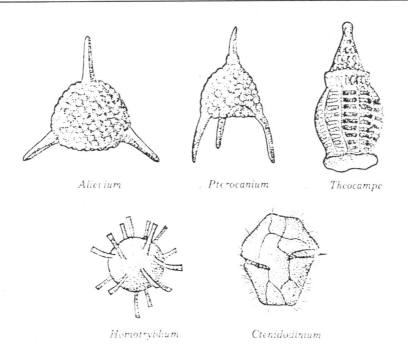

Alievium *Pterocanium* *Theocampe*

Hemotrybilum *Ctenidodinium*

Figure 13-3. (From *Treatise on Invertebrate Paleontology*, courtesy of The Geological Society of America and University of Kansas.)

GLOSSARY OF NEW TERMS

Tests: Hard part coverings or shells of forams.

Coquina: Grainy limestone made of fused shell fragments.

Oolite: Small, rare, sand-sized carbonate spheres or "snowballs" formed by wave action along carbonate beaches.

FURTHER READINGS

Davies, T.A. and D.S. Gorsline. Oceanic Sediments and Sedimentary Processes. *Chemical Oceanography*. J.P. Riley and R. Chester (editors), V.5. Orlando, Florida: Academic Press, 1976.

Heezen, B.C. and C.D. Hollister. *The Face of the Deep*. New York: Oxford University Press, 1971.

Moore, R.C. (editor). *Treatise on Invertebrate Paleontology*. Lawrence, KS: University of Kansas Press, 1964.

Pinet, P. *Oceanography: An Introduction to the Planet Oceanus*, West Publishing Company, 1992

Laboratory FOURTEEN

Point Counting

Statistical Evaluation of Beach Material

In past laboratories, we have carefully looked at the variety of materials that make up beaches. You have described sand from a plethora of sources and have picked, mounted and described foraminera and other microfossils from sand of several different ages. While doing so, you were asked to make estimations of proportions; the percentages of organic material versus inorganic materials, as well as the percentages of quartz, feldspar and other rock fragments. For the Timm's Point material, these observations are important and need to be considered when determining a variety of questions about the sand's origin and depositional environment. During this lab experience, you will progress beyond simple estimation and quantify your descriptions. You will also discover how much sampling is enough to obtain clear and reproducible results.

To accurately determine an exact distribution of any group, a complete count of all the material making up the whole must be done. This, of course, is highly impractical in many cases. This is especially true if one is working with materials with tremendous numbers of subsets, for example, the stars in the sky, the sand on a beach, or the voters in a national election. In lieu of counting each member of the whole, extrapolations from smaller subsets are made. There are several ways to approach the problem of sampling and each has its own advantages and limitations. Of course, we can not familiarize you with many of these methods in such a short time. We will therefore concentrate on only a few.

When a pollster is preparing to ask questions to project the opinions of the registered voters of the state of Illinois, she first considers several factors. Assuming that the questions are well prepared and the pollster can collect the information really desired; then most of the remaining considerations have to do with who to ask and how many. Their sample must be fair; people must be selected in an impartial way. Consideration must be given to factors seemingly unimportant. For instance, even a telephone poll may exclude some poor or transient peoples. Pollsters must be leery of selection bias; fortunately, rocks, shells and sand are not so difficult to deal with in this respect.

What size the poll should be is a difficult question. In the presidential poll of 1948, the Gallup poll used a sample size of 50,000 respondents which left an error of about 5%. Today, using advanced probability methods to select samples, they have been able to predict recent elections with surprising accuracy while sampling about 5 persons per 100,000! (Freedman et al, 1991). Just as it is impractical, if not impossible, to sample all 200 million eligible voters in the United States, it is equally impractical to sample all the grains on a beach or fish in the sea. The process in each of these examples is even more perplexing because voters from different geographic areas, different ethnic backgrounds, different economic circumstances, etc. have different political views. Different energy conditions along beach faces cause differences in size, density and therefore materials in the sand on the beach. The distribution of beach materials and the difficulty in taking truly random samples rules out a simple random sampling method for such studies.

Instead, like the pollsters, scientists frequently use a variation of simple random sampling known as multistage cluster sampling. This method is used where identifiable groups with known differences in composition of materials are delineated; then, within these groups a random sample is taken. In our beach example, the beach would probably be divided vertically from the shore face to the berm ridge and perhaps, laterally as well. This multistage method is one of the reasons that pollsters and scientists have been able to reduce sample size, yet increase the accuracy of their predictions or knowledge.

Another method to determine distributions of materials, such as those found on beaches, is by use of a grid or point count. To use this method, a few factors must be considered. The material must be distributed such that data will be useful when derived. Normally this means that there must be a homogeneous mix of the members sampled. In some cases, this is not necessary. For instance, when point counting is used in conjunction with multistage cluster sampling, the non-uniform distribution will be accounted for.

Grids are frequently set up over part of the sample area. In the studies of rock outcrops, geologists will often take a small unit such as a meter square randomly or not randomly chosen from the outcrop's surface and sample the entire area within its boundaries. Point counting is slightly different in that the entire surface in the grid boundary is not counted. Rather, one counts only those areas that appear under crossed lines within the grid that have been created in advance for that locality. Figure 14-1. Point counting can be somewhat more laborious than other forms of retrieving random data, but it is a method which insures that the researcher does not influence sample selection. Sometimes it is also possible to use tables of random numbers set up to help insure unbiased data gathering.

The size of the grid spacing used in point counting must be sufficient to "catch" all the subsets of interest. There is no ideal spacing to guarantee that all parts are sampled. Experimentation has shown that a grid with spacing approximately two times the length (longest axis) of the material being counted will provide good reproducible data.

It would be very cumbersome to have each of you crowd around a few microscopes to count sand grains. Instead, we are going to, in effect, enlarge the grains by several hundred times to make them visible to the naked eye and easily counted. We will be using various shells rather than sand or rocks. To sample them, you will fill cardboard flats so that at least one complete layer of material covers the bottom. The shells will come from a pile that we will assume is more or less homogeneous. There will still be a substantial pile of material left not sampled, after all the groups have filled their boxes. This "leftover" material will be thought of as the rest of the population or sample set.

Before you begin, there are a few concepts that need to be defined. There are several ways to draw information from studies as we will do. The most common methods are in the form of graphs. Yet even among graphs, there are several styles and types from which to choose. We will use a **histogram**. Histograms are one method to graphically represent discrete data that represents individual whole numbers or intervals of numbers with definite cut-off points. Histograms are drawn so that the frequency is shown on one axis and the interval on the other. Figure 14-1. The data is represented by rectangles or bars.

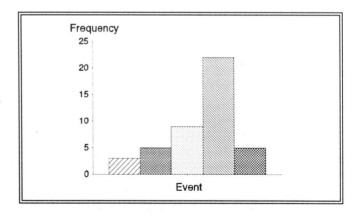

Figure 14-1. A Typical Histogram

Two terms used to determine how likely it is that our data is correct are **confidence interval** and **standard deviation**. Standard deviation is defined as the square root of the variance within the data collected. Practically speaking, two standard deviation units with their midpoint being the mean or average of the frequency, describes an area on a normal curve of data under which two thirds of the data points are found. Figure 14-2. In other words, if you have a point within one standard deviation unit from the average, that point will be among the two thirds majority of points. We could say then that we are nearly 67% (2/3 of the whole), certain that the true answer we were seeking (voters favoring one candidate, grains of one type, etc.) lies within the area of ± one standard deviation from the mean. In a sense then, this area defines a confidence interval of nearly 67%. But what if being only 67% confident is not good enough? Then we can adjust the level of confidence by expanding our range to include more data points. The range becomes less precise, but we become more confident that the real answer is included somewhere within that range.

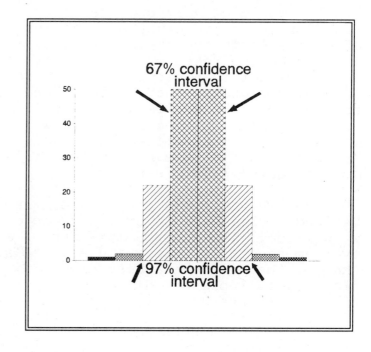

Figure 14-2. Area of one standard deviation on each side of average or mean. Area represents a 66.66% confidence interval. By increasing the confidence interval to 97%, the range gets larger but the precision decreases.

Dichotomous Keys

Dichotomous keys can be thought of as flow charts that ask questions of the user and, depending on their responses, direct the user further into the key until there are no more choices to be made. The correct answer is the only response left from this process of elimination. This form of identification key is the most frequently used for identifying trees, flowers, fish, shells and any number of other items.

You will create such a key during the exercises for today's lab. Several forms of dichotomous keys have been collected for you to view in the front of the classroom.

Exercises

1. **Creating a mechanism to sort unknowns.**
    As a class, use the prepared sample box in the front of the lab to sort the shells into categories. This box contains all the

shell types available in the material to be tested. You will first need to identify some form of reproducible categories. For our purposes, it does not matter what criteria you base your system of classification on. Size, composition, color, or shape are a few possibilities you may wish to consider. You should have at least five categories in your system.

2. **Criteria for classification of unknown objects.**
After sorting the materials into groups, describe each grouping in terms that any other student could readily understand. These should be such that they could reproduce the same clusters you did with some degree of confidence. This may not be as easy as it sounds.

Criteria for Categories
1)
2)
3)
4)
5)
6)
7)

3. **Creating a Dichotomous Key**
   Create a dichotomous key to your classification scheme from exercise #1. Remember, for this laboratory to be successful you must all use the same system of classification and be able to derive the same category for each sample specimen.

Dichotomous Key to Sorting Unknown Pile of "Enlarged Beach Sand"

4. **Point Counting**
   Fill a cardboard box with the materials from the "pile" such that the entire bottom is covered by a minimum of one layer of shells. (Of course, the "pile" may actually be in a bucket or bin in your lab.) Place the clear plastic sheet with a previously drawn grid over the box of shells. The grids were drawn prior to class in accordance with the recommendations for grid spacing described in the text. Look, from directly above the box, through the grid to the shells below. Record, according to your classification scheme, the category for each shell that appears directly beneath the intersection points on the grid. If the superimposed intersection falls exactly between two shells, then each shell needs to be counted as one half.

5. **Creating a histogram plot of data.**
   Plot your results on the histogram provided for trial one.

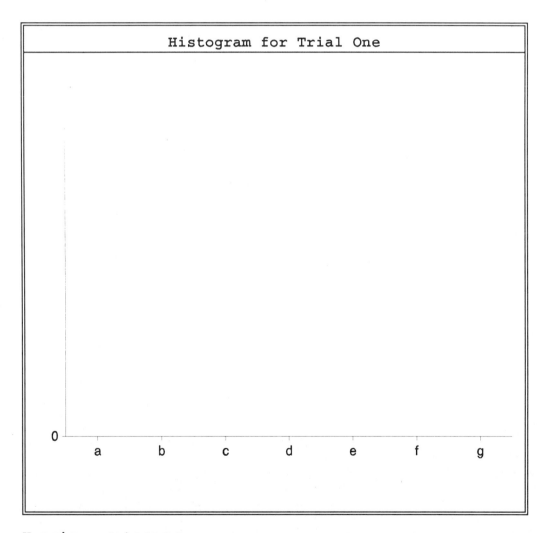

X axis = category
Y axis = frequency

6. **Reproducibility of results**
   Thoroughly mix the shells in your box, and repeat the process of recording all shells at the grid intersection. Plot the results from the second run on the histogram provided for trial two. Compare your results from trial one and trial two.

   X axis = category
   Y axis = frequency

7. **Calculation of confidence range**
   The next step in this lab may be the most worrisome for you if you have a phobia of numbers. This exercise is included to show you that the process of statistical evaluation, seen daily in opinion polls, newspaper diagrams, etc. is not magic. We will make a simple statistical evaluation to quantify the success of your tests.

   Compare the results of the entire class with yours. Do they seem (intuitively) to be acceptable? How can we be certain? How certain do we need to be? We are going to use a confidence interval of 95% for our discussion. Again, this means that we will be 95% confident that the true value lies in this range. If you were to score 95% on an exam, my guess is that you would be pretty confident. By selecting this (95%) interval in advance, we will be able to use a value in the equation that is predetermined for that confidence level. We can get the value to plug into the equation from tables set up just for that purpose. Although it may not seem correct at first, if we decrease the confidence interval to 90%, we actually have a more restrictive set of data points to work with. We will also need to have a value for the **standard deviation** of the material being sorted. For opinion pollsters, this is derived by past experience; for our purposes, it will be derived first by your instructors before class and modified after each lab group provides additional data.

The formula we will use is a standard for many statistical reviews (for 95% confidence intervals).

$$\mu = g \pm \frac{1.96 \times \textit{standard deviation for class}}{\sqrt{\textit{number of runs}}}$$

μ = the amount plus or minus that will be added to the class average to attain a 95% confidence interval (in the same way that a pollster says their results are accurate to ± so many percent)

g = Group average for each category
1.96 is the value derived from a Z-chart for the hypothetical standard normal curve which coincides with a confidence interval of 95%

Standard deviation for class will be provided.

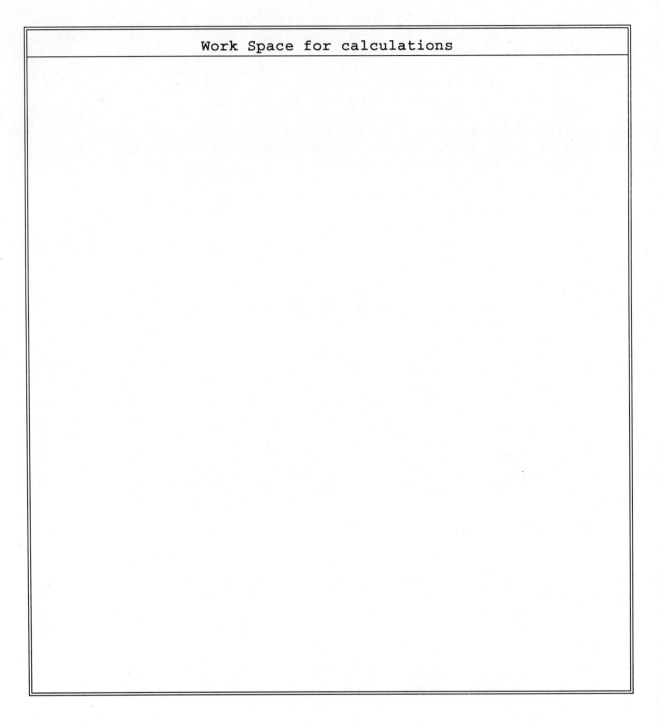

There is no perfect way to sample without including all the subsets in the sample. Every cluster of subsets has its statistical limits. We have calculated the tolerance (M, the number we add or subtract from our determined average) for a confidence level of 95%. This means that after we add or subtract our determined M value to our averages, we would still statistically be outside those limits one in every twenty sample sets. Most of the runs that would fall outside this 95% interval would be close to its limits. On rare occasions, however, there could be sample runs that would be quite different from our calculated confidence values. One way to accommodate these values outside the 95% interval is to expand the confidence interval to 96% or 97%. More of the extreme values would then fall into our acceptable range. But the range of acceptable data could increase to the point that almost any point would be permitted. This means that if we expanded to a 100% confidence interval, every possible combination of data sets from the "pile" would fall within our acceptable range. To know with 100% certainty, we would need to count every grain in the sample and the need to do this form of statistics would be eliminated. Yet, we already explained that this is impractical in many cases and the use of statistical evaluations is currently the only method we know of to deal with these large groups in a timely manner.

An important consideration in selecting a confidence interval is the ultimate purpose of the statistical evaluation in the first place. For instance, how confident do you believe pharmaceutical companies should be before a new drug is placed in the public market? Should the same level be needed before the shuttle lifts off? How about before a new discovery is made public about a more miserly carburetor? Each question being pondered has its own sets of limits that need to be addressed. Hopefully, you now have a better understanding to how those are derived.

GLOSSARY OF NEW TERMS

Histogram: One method to graphically represent frequency distributions drawn so that the frequency is shown on one axis and the interval on the other.

Confidence interval: The amount of assurance, usually represented by a percentage, that the results of data are not due to chance.

Standard deviation: The square root of the variance within the data collected.

FURTHER READINGS

Freedman, David, Robert Pisani, Roger Purves and Ani Adhikari. *Statistics*. 2nd ed. New York: W.W. Norton, 1991.

Friedman, Gary D. *Primer of Epidemiology*. New York: McGraw-Hill, 1980.

Pinet, P. *Oceanography: An Introduction to the Planet Oceanus*, West Publishing Company, 1992

Schopf, Thomas. *Paleoceanography*. Cambridge, MA: Harvard University Press, 1980.

> **I** *love the sea;*
> *she is my fellow-creature.*
> Francis Quarles, Emblems, 1635

Laboratory FIFTEEN

Caminalcules

Determining Biological Diversity and Phylogeny

Open nearly any biology or oceanography text, and you will read about the great diversity (different kinds) of organisms on the earth. Many estimates have been made regarding the total diversity of organisms on our planet. We know that about 1.5 million living plants and animals have been identified in the literature, and another 250,000 are mentioned in reports on the fossil record. Of course, these figures are much lower than the actual number. Some estimates range as high as 4 to 6 million modern living organisms currently living on our planet. Biologists are constantly discovering new creatures, especially within the phylum, Arthropoda, which contains the insects. Every expedition into the tropical rain forests yields tens of new sightings.

The reason for the small number of organisms from the earth's past is not for lack of effort. Paleontologists have not only the problem of discovery to deal with, but also the facts that:

- Fossilization is a rare event in itself usually requiring rapid burial found in few environments of deposition.

- Fossilization seldom occurs in terrestrial environments yet terrestrial organisms account for 80% of the known extant diversity on earth.

- Fossilization almost never preserves soft tissue; rather, hard parts like bones, teeth, logs or scales are preserved. Since most organisms do not have abundant hard parts, they have been left out of the record.

- Fossils weather and erode quickly and must be discovered soon after exposure.

- If discovered at all, fossils are usually found by someone who is not trained to recognize them as new finds or realize their importance.

The true diversity of extant organisms on earth is probably closer to 3 or 4 million. Paleontologists have estimated that the fossil record provides evidence for only about 5% of the probable total organisms that lived in the past. Hence, a rough and perhaps conservative estimate of 8 to 10 million species may have lived on this planet since the beginning of life on earth.

How can anyone make an accurate count of diversity? To do so at the species level, the only moderately well defined unit in the Linnaean classification scheme, one must understand what is meant by the term, "**species**". Biologists generally define the term as a collection of organisms that potentially share a common gene pool in nature through multiple generations.

Obviously, it is not practical to observe the mating habits, or results thereof, of every newly discovered organism. It is impossible in the fossil record. In general, the working definition of a species translates as: If organisms look like they could share a common gene pool successfully in nature, then they are considered as the same species. In other words, a species is a species if a competent expert says it is and other competent experts agree. In paleontology, this can only be done with respect to preserved hard parts and may be much more subjective than the designations made by biologists. However, as we will see, there are problems faced by both paleontologists and biologists in making taxonomic designations.

Certainly, two or more researchers could view the same organisms differently (and frequently do). The view accepted most frequently in subsequent references is considered correct until proof to the contrary arises or a new interpretation pervades. This form of majority rule may seem odd at first but has actually served the sciences well. Each discovery report is subject to the scrutiny of peer review. Of course there are problems in such a process and new interpretations and evaluations need to occur periodically as new data and methods of study are made available. However, over the scores of years that this method has been used, surprisingly few of these interpretations have been changed because the earlier ones were poorly done. If altered at all, they have been done so usually in light of new information.

Taxonomy is a far more complex and fascinating topic than we can discuss in a few pages here. We are going to allow you to find out for yourself exactly how this process works using an exercise that puts you in the role of the researcher making taxonomic decisions.

Exercises

Dr. Joseph H. Camin (1922 to 1979), from the Department of Entomology, University of Kansas, devised the "animals" used in this exercise. The imaginary animals, "caminalcules", were created in order to be used for testing certain theoretical aspects of taxonomy and phylogeny. We will use the caminalcules to introduce the concept and practice of phylogenies and taxonomies. In doing this exercise, you should begin to appreciate and acquire a better understanding about the complexities of evolutionary biology. The caminalcules are used here by permission of Emily Camin.

Two important points need to be made that are only tangential to the exercises:

- Diversity, the number of different kinds of organisms, does not equal abundance. Because an organism has great diversity does not mean it will be abundant, nor will abundant organisms be diverse.

- Although biologists have many advantages over paleontologists when determining the niche, habitat, biomechanics, and habits of modern organisms they work with, few of them have developed a working knowledge of those organisms beyond the history of a few generations. Working together, researchers can frequently help one another sort out potential problem points such as those found in convergence evolution where two creatures that look rather similar are actually not closely related at all. By charting the organism's "family tree," the differences in origins frequently become quite apparent.

1. **Creating a Caminalcule Phylogeny**
   Using the sheets of caminalcules provided for you, cut each caminalcule out and begin to build a **phylogeny** (family tree) showing the ancestor/descendant relationships between them. Note that each has been given a reference number and has been assigned to the correct time period from which it was discovered. Record your phylogeny on the ruled paper provided. Begin by placing the #66 in the center of the page within the area for the Triassic Period on the worksheet. Place #64 and #65 in the Jurassic Period's interval above but offset from #66. Because 64 and 65 must have descended from 66, draw a line across the period boundary, connecting 66 to 64 and 66 to 65. Connect each "caminalcule" by number in a like-wise manner summarizing your phylogeny (family tree) on the worksheet. You may find it helpful to set up a poster-sized board to manipulate the caminalcules better. Taping them down will keep them from blowing away.

Observe the following rules
○ All the caminalcules are descendent from the single Triassic form, #66.
○ No ancestor-descendent relationships exist among caminalcules of any single Mesozoic period or Cenozoic epoch, i.e. no horizontal connecting lines between species living at the same time can be drawn.
○ Each caminalcule in each time interval must have an ancestor in the immediately preceding period (or epoch).
○ Each branch in the phylogenetic tree must be a single branch with one exception of a double branch.
○ There are evolutionary dead ends.

2. **Classifying Caminalcules**

Assume that the caminalcules belong to a single class of organisms and that each specimen represents a different species. Construct a classification using circles to separate species belonging to a single higher taxonomic category. Each species must be in a genus and every genus incorporated into a family and so on. There can be mono-specific genera, a genus containing only one species or (theoretically) a single genus containing all 66 species of caminalcules. Normally, middle ground is found by most of the caminalcule researchers doing this exercise. However, even in the world of "real" biology, there are scientists who tend to be "lumpers" and others who tend to be "splitters".

Remember: the Linnaean Hierarchy

```
Kingdom
    Phylum
        Class
            Order
                Family
                    Genus
                        Species
```

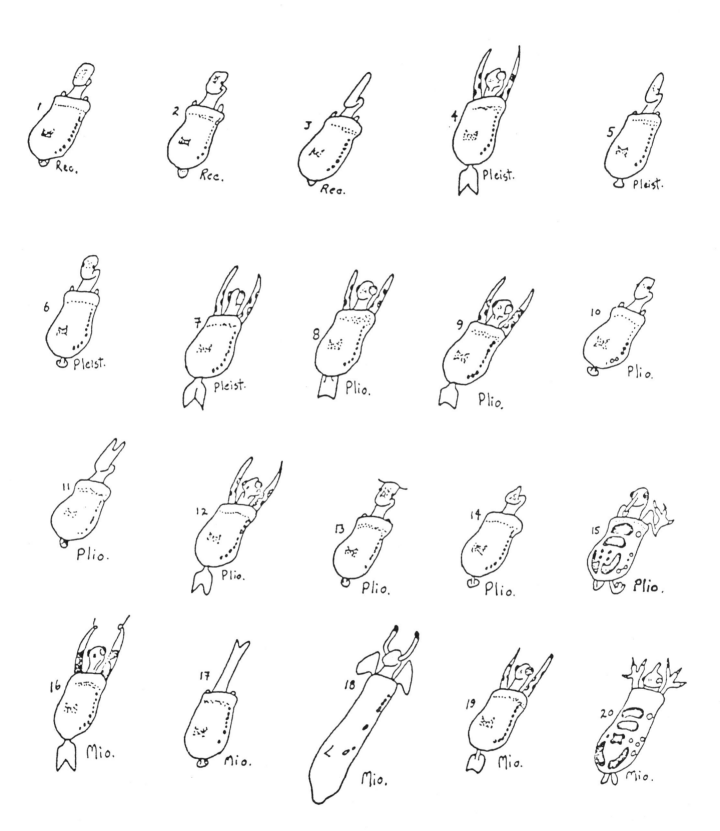

174

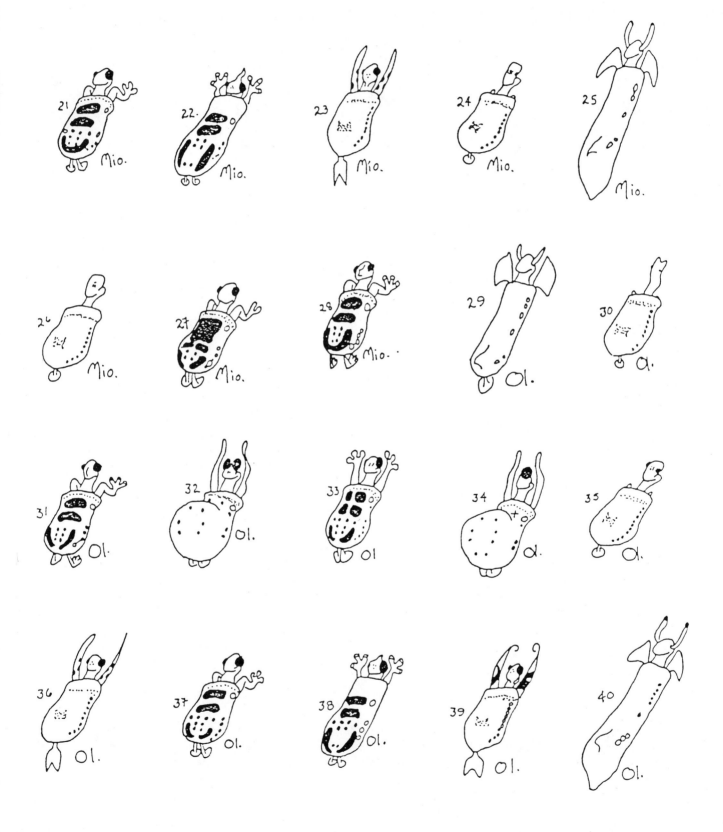

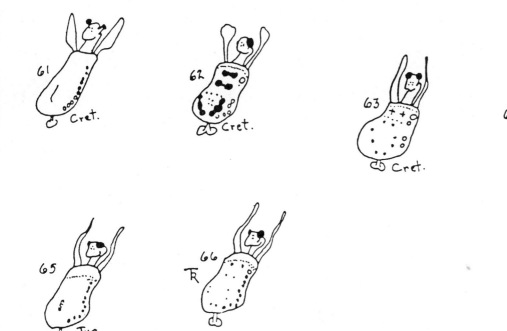

Caminalcule Phylogeny

RECENT	PLEISTOCENE	PLIOCENE	MIOCENE	OLIGOCENE	EOCENE	PALEOCENE	CRETACEOUS	JURASSIC	TRIASSIC

3. **Justifying phylogeny.**
Be able to justify the following:

   a. Your proposed phylogenetic arrangement of caminalcules.

   b. Your classification. Is it based on morphologic similarities or differences, on branching of the phylogenetic tree, on a time basis, or any combination of these? Do you tend to be a lumper or splitter? Which do you feel are the most important traits or parts of the caminalcules to be given priority when determining ancestor/descendant relationships? Why?

Answers to question #3
a.
b.

4. **Caminalcule ecology.**
Consider possible life habits and the evolution of the life habits in caminalcules.

How do you think this relates to the real world?

5. **Cladistical approach?**
Look up the term **cladistics**. Using the definition you find, does what you have done fit or not? How? Cladistics is but one of several models taxonomists use to create their phylogenies.

Answer to question #5.

GLOSSARY OF NEW TERMS

Species: A collection of organisms that potentially share a common gene pool in nature through multiple generations.

Phylogeny: Family tree.

FURTHER READINGS

Darwin, C. *On the Origin of Species by Means of Natural Selection, or the Preservation of Favored Races in the Struggle for Life.* London: John Murray, 1959.

May, R. M. 1988. How many species are there on Earth? *Science* 241(4872): 1441-48.

It is virtually impossible to keep out of the fresh air; this is perhaps the leading difference between life at sea and life on shore.

    W.J. Loftis, Orient Line Guide, 1885

Laboratory SIXTEEN

Invertebrate Ecology

Bivalve Ecology and Paleoecology

Geologists who specialize in studying fossils are called **paleontologists**; for the most part they are biologists of the ancient shallow seas. The rationale for making such a statement is that deposition occurs at a much greater rate in the oceans immediately adjacent to land than it does on land. And since rapid burial is one of the necessities for fossilization to occur, it makes sense that the phenomena of fossilization occurs most often in marine environments where instant burial is a common occurrence. Therefore, most of the information provided by the study of fossils is related closely to oceanography. While the most popular fossils may be of the great reptiles that roamed the earth over 100 million years ago (dinosaurs), these fossils actually account for a very small percentage of the total specimens studied by paleontologists. The dinosaurs, especially the big meat eaters (the rarest ones), just must have good press agents!

Fossils are often defined as "**any** indication of ancient life." This simple explanation will suffice for our needs, and is actually a very good definition. It means that an object need not be a "petrified" bone, shell, log, or tooth to be a fossil. Eggs, tracks, trails, borings, dens, coprolites (fossil feces), and gastroliths (gizzard stones) are all perfectly good and respectable fossils. You may be surprised to learn that these "trace" fossils can often tell us more about the organism's niche and environment than the remains of the organism itself. How can this be? Trace fossils are often interruptions of the earth, like dinosaur footprints in mud or a worm's burrow in the sand. If you try to move a freshly made track in mud today it is destroyed (unlike bones which can be carried great distances by the water in rivers or by the wind or can even be drug away by predators). Thus, with trace fossils we are certain at least that at one time the critter making the track had to have been at that place, and we know a little of the organism's condition - it was alive, alone, dragged its tail, walked with a limp, etc.

In today's lab we will view both trace and body fossils. We will, however, concentrate on those where the whole organism (hard parts) is preserved. Obviously, we cannot make a paleontologist out of you in one or two weeks, nor will we try. But if you can identify a fossil only to the class level, you may be able to infer some of its environmental parameters. This is especially true when the information from one fossil taxon is added to that from others found in the same strata at the same location. The information we will use for the fossils also holds true for many living forms as well.

We will be using several groups of fossils. Paleontologists refer to these groupings of organisms from the same rocks as **assemblages.** For this lab the assemblages will be marine and will contain mostly shells and exoskeletons of marine invertebrates. Each fossil taxon will be studied individually. Then the information for all of the taxa found together will be compared. Similarities or overlaps in the range of the environments suggested by the shell's morphology or form will be noted. We will attempt to determine an environment common to all the taxa studied. Thus, we will determine why these particular fossils were found together in the same rocks. Paleontologists borrow the terms for such studies from biologists: **Autecology**, the ecology of one type of organism, and **synecology,** the ecology of communities of organisms.

Before beginning your observations of the various groups, a review of a few biological principles may prove helpful. These principles are generalities; there are a few exceptions such as vestigial organs. (The exceptions will not be considered here.):

1) "Form follows function." This means that every part should have an evolutionary reason for being and is important.

2) "If you don't use it, you lose it." If the part had no purpose, it would be lost through the generations. This also implies that, if we are clever enough, we ought to be able to determine the function or use for each of the parts. Once we do, we may be clued into various environmental elements needed or desired by the organisms in question. Remember, however, the key words, "if we are clever enough." Often we don't have all the information needed, or we can be deceived into ignoring good clues and can draw the wrong conclusions. The rethinking of function is constant as new tools or ways of approaching this problem are discovered.

One very famous dinosaur group, the saurapods, like Dino on *The Flintstones*, was thought for many years to be restricted to living in swamps because the earliest reconstructions clearly showed that these brutes could not possibly support their own body weight. Today we believe those early assumptions were based on faulty reconstruction of the skeletons. We now believe that not only could they support their weight on all fours, but frequently stood on their hind legs to reach high into trees to feed in their forest (not swamp) homes.

If two groups of shells of the same taxon are compared, one group generally thick and the other thin, what physical reasons could there be for this shell structure difference? Could one be most suited to perhaps higher energy conditions? Which might be most likely to live in a higher energy environment? Generally, the thicker shells would protect the organism better in such a situation. If two sets of shells of the same density are observed, one group round like a marble and the other as flat as a fast food burger, are there evolutionary reasons why they were selected as preferred shapes? Which do you think would be most likely to live on a soupy, mucky substrate? What other characteristics can be treated by like comparisons within organisms of the same kind or between different types of critters? Size? Diversity? Abundance? Other shapes? Are there conclusions which might be drawn? These are the types of questions paleontologists and biologists routinely ask when trying to determine an organism's ecology, where it lived and how it "made its living." In today's lab we will do the same.

A powerful tool, yet limited in its use, is that of analogy. For our purposes this will mean comparing ancient fossils to recent close relatives. The assumption is that lifestyles in the past were probably similar to current lifestyles. Certainly there are exceptions, and the greater the time span between the lives of the organisms being studied, or the further apart the relationship in general, the more tenuous the analogy. However, many paleontologists and paleoceanographers say, "you must use the information available." Modern bivalve mollusks have well-defined niches that can, in part, be interpreted readily by inspection of their shells. If ancient organisms had the same traits seen in the modern counterparts, it is believed that they lived similar lives as the living ones. Some bivalve mollusks, like the modern pin shell *Pinna*, secure themselves to the strata by means of thread-like projections (byssus) that anchor the shell from excessive movement, yet still allow for some flexibility and movement in high energy situations. Some, like the misnamed "ship worm", *Torredo*, which is actually a mollusc closely related to the clams, can bore into rock or wood. Others, like common oysters, cement themselves directly to the hard bottoms or the shells of other organisms. Some like the "giant" clam, *Tridacna*, simply recline on the soft substrate. A few, like *Pecten*, can even move by swimming during a portion of their life cycle. However, since the great Permian extinction of invertebrates, most bivalve mollusks have turned away from living on or near the sediment-water interface to using their foot for burrowing instead. Some modern clams are found as deep as a meter below the surface. Figure 16-1.

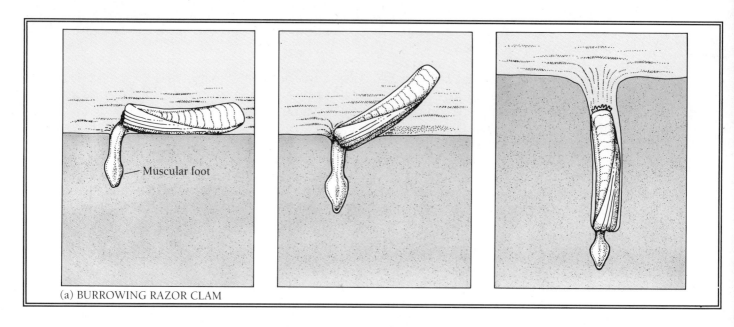

Figure 16-1. Burrowing bivalves. (From P. Pinet, *Oceanography: An Introduction to the Planet Oceanus,* **West Publishing Company, 1992.**)

This move into the substrate may have been made as a response to escape from predators capable of crushing the bivalve's shells. Along with the capability to burrow deeply, came the need to maintain contact with the sediment surface once buried. The bivalves still needed to respire and eat. This surface contact is done via siphons which carry nutrients to and wastes away from the organism. The burrowing bivalves retract the fleshy foot and siphons inside the shell when they are not being used or in times of stress. To do this requires storage space in the shell. In general, a correlation has been made between the size of this space, the paleal sinus, (Figure 16-2) and the length of the siphon. The length of the siphon is believed to directly correspond to the depth in which the organism routinely burrowed. This relationship is believed to exist in both ancient and modern bivalves.

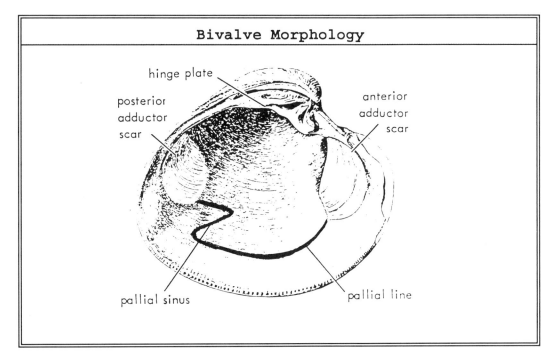

Figure 16-2. From *Treatise on Invertebrate Paleontology*, courtesy of The Geological Society of America and University of Kansas.)

Exercises

1. **Determining Bivalve Living Strategies**
    In the lab trays are 10 examples of bivalve mollusks both ancient (fossil) forms and modern examples. Examine each mollusc and determine if and how it lived on the substrate. Were they attached, free living, burrowing, boring or cemented? If they were burrowing forms, were they deep or shallow burrowers? The latter can only be determined on a relative basis by examining several specimens you believe are burrowers. Give a short rationale statement for your choice.

Bivalve Mollusc/Substrate Interaction Interpretations
1) _____ Evidence _____ _____
2) _____ Evidence _____ _____
3) _____ Evidence _____ _____
4) _____ Evidence _____ _____
5) _____ Evidence _____ _____
6) _____ Evidence _____ _____
7) _____ Evidence _____ _____
8) _____ Evidence _____ _____
9) _____ Evidence _____ _____
10) _____ Evidence _____ _____

2. **Determining Bivalve Habitats.**
   Using the shells in the tray provided, marked Part B, determine, as best as you can, the environment in which each of these bivalve mollusca lived. Be certain to address the following environmental factors - high or low energy, deep or shallow water, warm or cold, etc. You certainly may use information from Exercise A in your answers but concentrate on other indicators of environments as well. Prepare a list or narrative of your observations and the rationale for your decisions. Include a drawing for each key specimen used in making your decisions; label parts germane to your conclusions. Attach additional paper if needed.

Bivalve Habitat Interpretations

1) Environment _____
   Rationale _____
   _____
   _____

2) Environment _____
   Rationale _____
   _____
   _____

3) Environment _____
   Rationale _____
   _____
   _____

4) Environment _____
   Rationale _____
   _____
   _____

5) Environment _____
   Rationale _____
   _____
   _____

3. **Assemblage Interpretation**
   Using the information above and the same methodology, prepare similar lists and diagrams for two of the fossil assemblages in the lab trays. Then, determine the environments in which you believed they lived. Write your determinations and rationale for each clearly. Remember, this is <u>not</u> something that can be "looked up"; rather, it must be the result of careful reasoning. Feel free to consult other lab teams for their ideas. This is much the same procedure used by paleoecologists and paleoceanographers worldwide to interpret ancient conditions in the seas throughout time. Of course, other information can be used: data derived from stratigraphy (the layers of the rock or

sediment), sedimentology (the disturbances in the deposition process and how the sediment arrived where it is now found), trace fossil analysis, paleoclimatology, and any other fields which may shed light on the area being researched.

Assemblage Interpretations

Fossil Assemblage # _____
     Environment _____
     Rationale _____
 _____
 _____

Fossil Assemblage # _____
     Environment _____
     Rationale _____
 _____
 _____

GLOSSARY OF NEW TERMS

Fossil: Any indication of ancient life.

Autecology: The ecology of a single kind of organism.

Synecology: The ecology of organisms communities.

Assemblage: Groups of fossils collected from the same rock layer at one locality.

Paleontologist: Geologists that specialize in studying ancient life.

FURTHER READINGS

Cushing, D.H., and J.J. Walsh. *The Ecology of the Seas.* Philadelphia: Saunders, 1976.

Hedepeth, J.W. (ed). Treatise on Marine Ecology and Paleoecology, *Ecology.* Geological Society of America, 1: Memoir 67, 1957.

Isaacs, J.D. The nature of oceanic life. *Scientific American,* 221: 65-79, 1969.

Levinton, J.S. *Marine Ecology.* Englewood Cliffs, NJ: Prentice-Hall, 1982.

Longhurst, A., and D. Pauly. *Ecology of Tropical Oceans.* London: Academic Press, 1987.

Parsons, T.R., M. Takahashi, and B. Hargrave. *Biological Oceanographic Processes,* 3rd ed. Oxford: Pergamon Press, 1985.

Pinet, P. *Oceanography: An Introduction to the Planet Oceanus,* West Publishing Company, 1992

Sumich, J.L. *An introduction to the biology of marine life.* Dubuque, Iowa: Wm. C. Brown, 1976.

Thorson, G. *Life in the Sea.* New York: McGraw-Hill, 1971.

"But the line between the natural and the preternatural is very cloudy. Natural things occur, and for most of them there's a logical explanation. But for a whole lot of things there's just no good or sensible answer... In a way, sharks are like tornadoes. They touch down here, but not there. They wipe out this house but suddenly veer away and miss the house next door. The guy in the house that's wiped out says, 'Why me?' The guy in the house that's missed says, 'Thank God.'"

Matt Hooper, Ichtheologist in Peter Benchley's Jaws, 1974

Laboratory SEVENTEEN

Marine Ecology

Fish Aquarium Study

Marine ecology may be the most glamorous of the oceanographic sub-disciplines. Nearly all of us, at one time or another, have dreamed of being a porpoise trainer or swimming in the seas alongside a giant whale. At the very least, most of us have rooted for the good guy in the "Jaws" sequels. To study marine ecology, ideally, we would all put on scuba equipment, go down to the depths of the ocean, and watch our chosen community continue on in its day-to-day existence. We could become unobtrusive observers chronicling the autecology of each critter. The scuba method is not as ideal or problem-free as one might first believe from either a logistical or a scientific standpoint. Observation time is severely limited by air supply, and, among other problems, it is difficult to document that divers are "unobtrusive observers" who have not affected the behavior of the community members they are observing. We are, therefore, going to bring the fish and other marine invertebrates to us instead of going to the fish.

Many biology departments maintain salt-water aquariums of substantial size. The aquarium community often includes normal, tropical-subtropical, shallow-water, reef-dwelling organisms. A self-contained, self-sustaining tank would be optimal, but the size required for such a system prohibits this. No such tank could be brought in for our observations. Even the large showplace tanks in the world's major aquariums, such as the Shedd Aquarium in Chicago, must feed the system. Just as in the aquariums you see in people's homes or in the dentist's office, the fish are not supposed to eat one another because it gets too expensive to replace the fish all the time. In these aquariums, the lower portion of the food web is artificially supplied by the care takers in sufficient quantities so that upper level feeders remain well fed. Increased populations by birth are removed and only allowed on a replacement basis.

The aquarium you will be observing is different because it was designed to let a simple food chain develop within it. Figure 17-1. This system is not large enough to sustain a complex food web. In the aquarium most of you will observe, we expect the organisms to live as closely as possible to the way they would in the wild. This marine system includes plants, primary consumers, secondary consumers and a top carnivore as well.

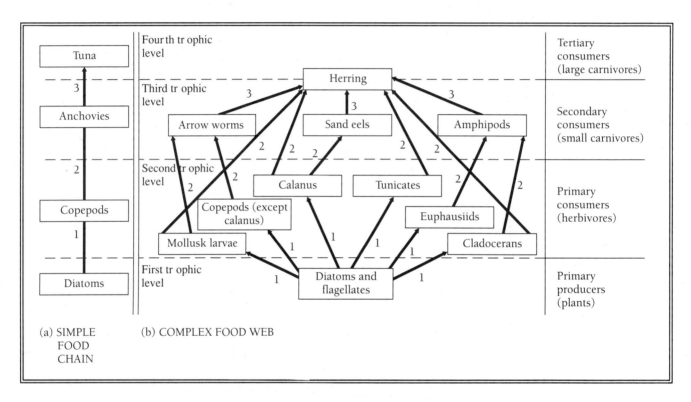

Figure 17-1. Trophic relationships. A) A simple marine food chain, B) A complex marine food web more closely approximates conditions in the sea. (From P. Pinet, *Oceanography: An Introduction to the Planet Oceanus*, West Publishing Company, 1992.)

If a salt water aquarium, designed with food chains intact, is not available, do not despair. Although still another step away from ideal conditions, a fresh water home aquarium, or one in a friend's room, will do. As any good aquarist knows, in even the simplest aquarium set, care must be made to include only fish compatible with one another. In addition to having fish that are hospitable or tolerant of each other, it is best to have a diverse population as well. This will include schooling as well as solitary fish, those that prefer the near-surface condition and those that are bottom dwellers. Most good books for the hobbyist fish keeper will give some insight as to the preferences of fish readily available at the neighborhood pet stores. Keeping the various traits of the fish in mind will not only enhance the appearance of the aquarium but also allow for a much healthier community to develop.

Exercises

1. **Aquarium Observation**

    a. List the kinds (class level systematics will do fine) of organisms present on the day you observe the aquarium.

Organisms in the aquarium

    b. Choose one type of organism to watch carefully for an extended period of time - at least 1/2 hour. Keep track of the movement and other activities of your chosen organism, especially interactions with both alike and different organisms. Write your observations on a separate page.

c. Can you make **any** inferences about your critter concerning preferences in water depth, energy conditions, what they eat, what eats them? Are there other organisms they seem to avoid? Do they interact with their own kind? Do they avoid any organisms? Are they territorial?

Organisms preferences

d. In the library, look up material on the niche and habitat of the organism you observed and determine how it lives in its native element. Discuss any differences in what you read from what you observed.

Differences between native and aquarium habitat

2. **Analyzing Study**
This is the most important question of the exercise. The merits, or lack thereof, of aquarium studies have long been debated. Discuss limits to this exercise. What can or cannot be learned by aquarium studies? Aquarium studies cannot be all bad since ichthyologists still use aquarium studies in recently published reports. Give this question careful thought. It will be discussed in class.

Limitations & attributes of aquarium studies

FURTHER READINGS

Cushing, D.H., and J.J. Walsh. *The Ecology of the Seas.* Philadelphia: Saunders, 1976.

Hedepeth, J.W. (ed). Treatise on Marine Ecology and Paleoecology. *Ecology.* Geological Society of America. 1: Memoir 67, 1957.

Levinton, J.S. *Marine Ecology.* Englewood Cliffs, NJ: Prentice-Hall, 1982.

Longhurst, A., and D. Pauly. *Ecology of Tropical Oceans.* London: Academic Press, 1987.

Parsons, T.R., M. Takahashi, and B. Hargrave. *Biological Oceanographic Processes*, 3rd ed. Oxford: Pergamon Press, 1985.

Pinet, P. *Oceanography: An Introduction to the Planet Oceanus*, West Publishing Company, 1992

He who commands the sea has command of everything.

Cicero, Epistolae ad Atticum X

TRITON'S REVENGE

Who Rules The Seas?

A Simulation on the Law of the Sea Conference

It has long been recognized that to protect national interests and establish world stability, laws must govern commercial and military activities beyond the terrestrial borders of countries. Now the exploitation of the seas has been complicated further by problems such as the polluting of the oceans. It is imperative to develop international agreement on the jurisdiction of nations and an understanding of the rights and responsibilities of all parties beyond these agreed limits.

In 1609, Dutch jurist, Hugo Grotius, established the concept of doctrine that the seas should be free to all nations. Problems in this doctrine were many but were substantially reduced. In 1702, when Cornelius Van Bynkershonk published a report, which was generally agreed to, allowing for national sovereignty to extend to a distance protected by shore-based cannons. One can see that this was only a temporary solution, especially in light of the land based projectiles now capable of striking anywhere in the world!

In 1958 the first United Nations Conference on the Law of the Sea was convened in Switzerland and produced the first set of modern limits of control of seas near a country's borders. The conference used the concept of "continental shelf" as the principle definer of national jurisdiction. Unfortunately, at that time, the concept of continental shelf was not well defined or agreed upon.

A series of meetings on the Law of the Sea have been held nearly every year subsequent to the early 1970's. In 1982, 130 of 151 countries present agreed to a new Law of the Sea Treaty. Only four countries openly opposed the signing: The United States, Turkey, Israel and Venezuela. Each felt private corporations best suited to

develop the mining of the sea floor, might be penalized to the point that their ventures would be unprofitable, and thus abandoned if this version of the Law of the Sea Treaty was ratified. Many of the abstaining votes were by countries with well developed technologies capable of exploiting various resources of
the seas.

The scenario in the simulation "game" below is far from trite. Several noted scholars have predicted that a situation capable of pivoting the world toward global conflict is only one colossal mineral or energy discovery away. A "gold rush" in the ocean can easily be envisioned as well as the disaster such an event could cause. If such a disaster did occur, who would be the next Wyatt Earp?

Each of you will play the role of an ambassador from one of the countries assembled for a new Law of the Sea Conference. While the simulation is progressing, you must remain true to the spirit of your country's position as explained in a handout you will receive from your instructor. However, you and your fellow ambassadors should consider that you are citizens of the entire world as well as citizens of your own state. Advantages gained for the state at the expense of world order may be meaningless.

Your instructor will provide more information and the needed handouts which, for reasons you will soon appreciate, are not printed in this edition. A brief scenario and description of the participants follows.

TRITON'S REVENGE- WHO RULES THE SEA?
SCENARIO

TIME: Present

SITUATION: All prior treaties and laws of the sea have collapsed. Throughout the world a sense of greed and "might is right" prevails. Many technologically superior nations are planning programs to exploit precious minerals, energy and food resources from the world's oceans. The United Nations feels that the only potential for a negotiable solution capable of averting conflict rests in their organization. In a last ditch effort, the United Nations has selected six representative countries from a cross-section of its membership to attempt a compromise in the form of a new "Law of the Sea". These six representative countries must develop a solution. The entire world is concerned that, if this conference is not a success, world war may be eminent as several countries are poised to set their plans into operation.

A press conference has already been arranged for a time immediately following the conference to announce the results of the negotiations to an expectant world. Conferees are to draft the major concepts of their plan, if one is agreed upon. Details will be worked out later in subsequent negotiations; however, time is of the essence.

Conferees must address the following topics:

o   How far from any country's boundaries can militarily defensible borders extend?

o   How far can any country extend its economic boundaries?

o   What happens when two country's defensible or economical boundaries overlap?

o   Since some countries are more technologically advanced than others, should the oceans be free economic zones for any country or corporation to manage as its resources allow, or should the information and profits from the ocean's exploitation somehow be divided? If so, how could this be accomplished (i.e. to whom and how much)?

o   Who will settle disputes when they arise?

The countries that have been selected to attend these problems are: **Eastern Slobovia, Pofolksia, Tunarus, Upper Crust, Wannabe and Gobland.** Descriptions of the countries can be found on page 200.

BRIEFS ON COUNTRIES AT THE LAW OF THE SEA CONFERENCE

Eastern Slobovia, a land-locked country with a single river deep and wide enough to be used for a port to ship produce grown in the country. The river is the natural border between two other countries, neither of which are at the conference. Eastern Slobovia has no navy, but a sizable army and is considered a stabilizing force in its region. The government is a monarchy. Its medium-sized population is well educated. The royal family has been in power for centuries.

Pofolksia, a land-locked country that is very poor. The citizens of Pofolksia frequently live on the edge of poverty. This is a recently formed democratic country. They have no military strength besides a nominal internal police force.

Tunarus, a small island country that depends on fishing for its livelihood. Tunarus has a small military used mostly to patrol its fishing grounds. Tunarus has warned countries that they consider a two-hundred mile fishing zone around the island as a part of their sovereignty.

Wannabe, a large country with few good ports to support its under-developed, principally agricultural and limited industrial economy. The country has a great military including a powerful navy.

Upper Crust, a medium-sized island country that has a powerful navy. It currently has technical superiority to most other world powers.

Gobland, a developing country with a coastline of ports but no navy to speak of. The country has a lot of under-developed resources including precious metals, oil and gas, and tremendous fishing grounds found just off its coast. Gobland has a history of military seizures of unstable governments.

TALLY SHEET TO SUMMARIZE DATA

	Eastern Slobovia	Pofolksia	Tunarus	Wannabe	Upper Crust	Gobland
Military Borders						
Economic Boundaries						
Overlap Boundaries						
Profits Divided						
Settles Disputes						

	Eastern Slobovia	Pofolksia	Tunarus	Wannabe	Upper Crust	Gobland
Military Borders						
Economic Boundaries						
Overlap Boundaries						
Profits Divided						
Settles Disputes						